Approximate Arithmetic Circuit Architectures
for FPGA-based Systems

Salim Ullah • Akash Kumar

Approximate Arithmetic Circuit Architectures for FPGA-based Systems

 Springer

Salim Ullah
TU Dresden
Dresden, Germany

Akash Kumar
TU Dresden
Dresden, Germany

ISBN 978-3-031-21296-3   ISBN 978-3-031-21294-9   (eBook)
https://doi.org/10.1007/978-3-031-21294-9

This Springer imprint is published by the registered company Springer Nature Switzerland AG
The registered company address is: Gewerbestrasse 11, 6330 Cham, Switzerland

Dedicated to our families and all those who always give it one more try

Preface

From the initial computing machines, Colossus of 1943 and ENIAC of 1945, to modern high-performance data centers and Internet of Things (IOTs), four design goals, i.e., high-performance, energy-efficiency, resource utilization, and ease of programmability, have remained a beacon of development for the computing industry. During this period, the computing industry has exploited the advantages of technology scaling and microarchitectural enhancements to achieve these goals. However, with the end of Dennard scaling, these techniques have diminishing energy and performance advantages. Therefore, it is necessary to explore alternative techniques for satisfying the computational and energy requirements of modern applications. Towards this end, one promising technique is analyzing and surrendering the strict notion of correctness in various layers of the computation stack. Most modern applications across the computing spectrum—from data centers to IoTs—interact and analyze real-world data and take decisions accordingly. These applications are broadly classified as Recognition, Mining, and Synthesis (RMS). Instead of producing a single golden answer, these applications produce several feasible answers. These applications possess an inherent error-resilience to the inexactness of processed data and corresponding operations. Utilizing these applications' inherent error-resilience, the paradigm of approximate computing relaxes the strict notion of computation correctness to realize high-performance and energy-efficient systems with acceptable quality outputs.

The prior works on circuit-level approximations have mainly focused on Application-specific Integrated Circuits (ASICs). However, Application-specific Integrated Circuit (ASIC)-based solutions suffer from long time-to-market and high-cost developing cycles. These limitations of ASICs can be overcome by utilizing the reconfigurable nature of Field Programmable Gate Arrays (FPGAs). However, due to architectural differences between ASICs and Field Programmable Gate Arrays (FPGAs), the utilization of ASIC-based approximation techniques for FPGA-based systems does not result in proportional performance and energy gains. Therefore, to exploit the principles of approximate computing for FPGA-based hardware accelerators for error-resilient applications, FPGA-optimized approximation techniques are required. Further, most state-of-the-art approximate arithmetic operators do not have a generic approximation methodology to implement new approximate designs for an application's changing accuracy and performance requirements. These works also lack a methodology where a machine learning model can be used to correlate an approximate operator with its impact on the output quality of an application. This book focuses on these research challenges by designing and exploring FPGA-optimized logic-based approximate arithmetic operators. As multiplication operation is one of the computationally complex and most frequently used arithmetic operations in various modern applications, such as Artificial Neural Networks (ANNs), we have, therefore, considered it for most of the proposed approximation techniques in this book.

The primary focus of the work is to provide a framework for generating FPGA-optimized approximate arithmetic operators and efficient techniques to explore approximate operators for implementing hardware accelerators for error-resilient applications. Towards this end, we first present various designs of resource-optimized, high-performance, and energy-efficient accurate multipliers. Although modern FPGAs host high-performance Digital Signal Processing (DSP) blocks to perform multiplication and other arithmetic operations, our analysis and results show that the orthogonal approach of having resource-efficient and high-performance multipliers is necessary for implementing high-performance accelerators. Due to the differences in the type of data processed by various applications, the book presents individual designs for unsigned, signed, and constant multipliers. Compared to the multiplier IPs provided by the FPGA Synthesis tool, our proposed designs provide significant performance gains. We then explore the designed accurate multipliers and provide a library of approximate unsigned/signed multipliers. The proposed approximations target the reduction in the total utilized resources, critical path delay, and energy consumption of the multipliers. We have explored various statistical error metrics to characterize the approximation-induced accuracy degradation of the approximate multipliers. We have also utilized the designed multipliers in various error-resilient applications to evaluate their impact on applications' output quality and performance.

Based on our analysis of the designed approximate multipliers, we identify the need for a framework to design application-specific approximate arithmetic operators. An application-specific approximate arithmetic operator intends to implement only the logic that can satisfy the application's overall output accuracy and performance constraints. Towards this end, we present a generic design methodology for implementing FPGA-based application-specific approximate arithmetic operators from their accurate implementations according to the applications' accuracy and performance requirements. In this regard, we utilize various machine learning models to identify feasible approximate arithmetic configurations for various applications. We also utilize different machine learning models and optimization techniques to efficiently explore the large design space of individual operators and their utilization in various applications. In this book, we have used the proposed methodology to design approximate adders and multipliers.

This book also explores other layers of the computation stack (*cross-layer*) for possible approximations to satisfy an application's accuracy and performance requirements. Towards this end, we present a framework to allow the intelligent exploration and highly accurate identification of the feasible design points in the large design space enabled by *cross-layer* approximations. The proposed framework utilizes a novel Polynomial Regression (PR)-based method to model approximate arithmetic operators. The PR-based representation enables machine learning models to better correlate an approximate operator's coefficients with their impact on an application's output quality.

Dresden, Germany Salim Ullah
Dresden, Germany Akash Kumar
September 2022

Acknowledgments

We would like to thank our group members and collaborators, especially Dr. Siva Satyendra Sahoo, for their continued support in realizing this work.

Contents

Acronyms

AI	Artificial Intelligence.
ALM	Adaptive Logic Module.
ANN	Artificial Neural Network.
ASIC	Application-specific Integrated Circuit.
BO	Bayesian Optimization.
CGP	Cartesian Genetic Programming.
CLB	Configurable Logic Block.
CNN	Convolutional Neural Network.
CPD	Critical Path Delay.
DNN	Deep Neural Network.
DoF	Degree of Freedom.
DSE	Design Space Exploration.
DSP	Digital Signal Processing.
DTR	Decision Tree Regression.
DVFS	Dynamic Voltage and Frequency Scaling.
ECG	Electrocardiographic.
EPF	Evaluated Pareto Front.
FE	Fidelity Error.
FIR	Finite Impulse Response.
FPGA	Field Programmable Gate Array.
GA	Genetic Algorithm.
GBR	Gradient Boosting Regression.
GOPS	Giga Operations per Second.
GPU	Graphics Processing Unit.
GS	Gaussian Smoothing.
HDL	Hardware Description Language.
HLS	High-level Synthesis.
IC	Integrated Circuit.
IP	Intellectual Property.
LUT	Lookup Table.
MAC	Multiply-Accumulate.

MAE   Mean Absolute Error.
MBO   Multi-objective Bayesian Optimization.
ML   Machine Learning.
MLP   Multi-layer Perceptron.
MOEA   Multi-objective Evolutionary Algorithm.
MSE   Mean Square Error.
NPU   Neural Processing Unit.
PDP   Power-Delay Product.
PP   Partial Product.
PPA   Power, Performance and Area.
PPF   Predicted Pareto Front.
PPP   Processed Partial Product.
PR   Polynomial Regression.
PSNR   Peak Signal-to-noise Ratio.
RCA   Ripple Carry Adder.
RFR   Random Forest Regression.
RL   Reinforcement Learning.
RMS   Recognition, Mining, and Synthesis.
SGD   Stochastic Gradient Descent.
SIMD   single-instruction-multiple-data.
SP   Sub-product.
SSIM   Structural Similarity Index.
SVR   Support Vector Regression.
WER   Word Error Rate.

Chapter 1
Introduction

1.1 Introduction

Over the past few decades, the computing industry has witnessed a rapid growth in the computational demands of applications from various domains. The semiconductor industry has remained successful in satisfying the computational needs of the market by regularly producing Integrated Circuits (ICs) with better performance than the previous generations. According to Moore's law, the technological improvements in each successive generation result in fabricating an almost double number of transistors on the same chip size and hence increasing the operating frequency of the chip [1]. However, with the recent advances in technology scaling (5 nm node), as the transistor size approaches the atomic scale, it becomes harder and more expensive to double the number of transistors as predicted by Moore's law [2]. Similarly, Dennard scaling states that despite fabricating more transistors on the chip, the power density of the chip remains constant as the operating voltage and current scale down [3]. The chips' constant power density had encouraged chip manufacturers, until recently, to exploit higher operating frequencies to achieve higher performance/watt. However, at a smaller feature size (<65 nm node), the aggregate leakage current cannot be ignored, and it can result in the thermal runaway of the chip [4]. To provide high-performance systems, computing industry moved toward multicore systems. However, even with multicore systems, not all cores can be operated at maximum power all the time. This phenomenon of power-gating some cores to reduce the overall power consumption of the chip is commonly known as dark silicon [5]. Furthermore, despite employing multiple cores, many applications exhibit limited performance improvements due to various issues related to parallel execution, such as dependencies among operations, serial computation, synchronization among various operations, and memory bandwidth bottleneck [6].

A promising solution to provide energy-efficient and high-performance computing systems is to utilize dedicated accelerators for various tasks. For example, the Apple A14 Bionic chip hosts two 64-bit high-performance cores and four 64-

bit energy-efficient cores. It also provides a dedicated graphics processing unit, an image signal processor, a 16-core neural network accelerator, and a matrix scalar multiplication accelerator [7]. Compared to a general-purpose processor, the application-specific implementations provide better performance in terms of resource utilization, execution time, and energy efficiency.

It is interesting to observe that during this era of technology and architectural innovations, the nature of the computing workload has also changed significantly over the past few years. For example, the computing workload at data centers is mainly related to analyzing a large volume of digital data and making intelligent decisions accordingly. Similarly, on the other end of the computing spectrum, mobile devices and embedded systems at the edge interact with real-world data and provide a real-time experience to the users. A common attribute of these diverse and modern applications is the ability to produce multiple feasible outputs instead of a single golden output. For example, a search engine can return multiple feasible choices instead of a single golden answer while searching for the best sushi restaurant in town. Similarly, an audio or video processing application on a mobile device can use various encoders to produce different acceptable quality outputs. Applications capable of producing multiple feasible outputs, instead of a golden answer, possess an inherent error resilience to the inaccuracies in their intermediate computations. The inherent error resilience provides another promising venue—*the paradigm of approximate computing*—for implementing resource-efficient, high-performance, and energy-efficient accelerators for a wide range of applications [8]. The approximate computing paradigm introduces deliberate approximations across various layers of the computation stack to provide an accuracy-performance trade-off and reduce the overall computational complexity, required computational resources, execution time, and the energy requirement of an application [9]. For example, the compute-intensive single-precision floating-point multiplications in an Artificial Neural Network (ANN) accelerator can be replaced by an 8-bit fixed-point multiplications to reduce the overall computational complexity of the operation by an allowed degradation in the generated output quality.

Among the various layers of approximation, architecture- and circuit-level techniques have been a major focus of research for resource-constrained embedded systems [10–14]. However, these techniques have mainly focused on Application-specific Integrated Circuits (ASICs)-based systems. This book mainly concentrates on designing accurate and approximate arithmetic blocks optimized for Field Programmable Gate Arrays (FPGAs)-based systems. Further, it also explores other layers of the computation stack for possible approximations. Toward this end, it presents a framework for the joint analysis of approximations across multiple layers of the computation stack.

The rest of the chapter is organized as follows. Section 1.2 defines the inherent error resilience of applications and presents an overview of the various sources contributing to application error resilience. Section 1.3 introduces the approximate computing paradigm for error-resilient applications. It also provides a summary of some of the state-of-the-art approximate computing works targeting different layers of the computation stack. Section 1.4 introduces the research problems investigated in this book. It also elaborates on the various associated research challenges examined in this work. Section 1.5 provides an overview of the flow of the book and how the various chapters are correlated. Finally, Sect. 1.6 summarizes the novel contributions presented in this book and a description of the relevant peer-reviewed publications on which the contributions are based.

1.2   Inherent Error Resilience of Applications

Inherent error resilience of an application refers to its ability to produce acceptable quality results despite some inaccuracies (approximations) in its underlying operations and utilized data. The term "acceptable quality results" refers to the fact that these applications produce a range of different quality outputs instead of producing a single precise answer. Therefore, it is essential to define application-specific output evaluation metrics to characterize the quality of the generated outputs. For example, for an image processing application, the output quality of a processed image is usually measured using the Peak Signal-to-noise Ratio (PSNR) metric. Similarly, for measuring the output accuracy of a speech recognition system, the Word Error Rate (WER) is a commonly used metric. Error-resilient applications cover multiple domains of applications, e.g., digital signal processing, audio and image processing, machine learning models, data communication protocols, and applications belonging to the Recognition, Mining, and Synthesis (RMS) field. It should be observed that the error resilience of an application does not signify that all underlying computations and associated data can tolerate inaccuracies. Some parts of the applications, such as the control flow, are sensitive to accuracy and should be executed with the highest available accuracy and precision. The inaccuracies (errors, approximations) in the accuracy-sensitive parts of an application can have catastrophic results. For example, the utilization of 16-bit and 32-bit precisions[1] is considered the main reason for the destruction of the *Ariane 5* rocket in 1996 and the *Boeing 787* aircraft in 2013, respectively [15]. Some recent works, such as [8, 16, 17], have proposed different frameworks to analyze an application and identify the error-resilient and sensitive parts of an application. For example, the framework presented in [8] has analyzed 12 different applications from the RMS domain and identified that these applications spent on average 83% of their execution time in error-tolerant operations. Further, as shown in [18], the error-tolerant elements of applications are, on average, the main contributors to the overall resource utilization and energy consumption of the applications.

As described in Fig. 1.1, several factors contribute to the error resilience of these applications [8]. These factors can be grouped into the following categories:

---

[1] Precision scaling is a commonly explored technique in the paradigm of approximate computing for error-resilient applications.

Fig. 1.1 Sources of inherent error resilience of applications

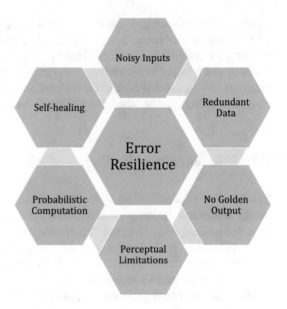

- **Inputs**: Many applications process real-world noisy and redundant data. For example, intelligent virtual assistants, such as Alexa, Siri, and Google Assistant, process voice commands to answer different users' queries. These voice commands often contain a large amount of noise and carry redundant information. The corresponding algorithms to process such type of input data are robust to small errors and variations in the input. Further, many applications process similar noisy data multiple times, which also adds to the overall error resilience of the applications.
- **Outputs**: Many applications produce multiple outputs instead of a single golden answer. For example, a search engine usually returns multiple feasible answers matching the provided search string. The absence of the constraint to produce a single precise answer—golden answer—allows the applications to tolerate inaccuracies (approximations) in the input data and underlying operations. Similarly, the perceptual limitations of human beings also acknowledge the generation of various quality outputs. For example, Fig. 1.2 shows three different processed versions of the same logo of Technische Universität Dresden. The logos presented in Fig. 1.2b and c have lower resolutions and also require less storage spaces than the logo in Fig. 1.2a. However, a human eye can hardly find any visual difference between Fig. 1.2a and b. Perceptual limitations allow the applications to generate multiple acceptable quality outputs and manifest error resilience.
- **Computations**: Applications such as Bayesian inference and K-means clustering employ probabilistic computational models. Probabilistic computational models provide estimates and are inherently error resilient. Furthermore, applications like ANNs utilize an iterative convergence approach to refine the various

(a)   (b)   (c)

Fig. 1.2 Perceptual limitations for error resilience of applications: TUD logo (**a**) 792 × 238 and 182.1 kB, (**b**) 198 × 60 and 19.3 kB, (**c**) 79 × 23 and 4.5 kB

parameters of the application. The iterative execution of the processes results in the attenuation of various inaccuracies in the data and operations. For example, many works such as [19, 20] exploit this feature to train ANNs with low bit-width quantization schemes.

1.3 Approximate Computing Paradigm

The approximate computing paradigm leverages the error resilience of applications to deliberately reduce the precision of the data and underlying operations to improve the application's overall performance. It should be noted that the approximate computing paradigm is different from two related computing techniques, i.e., error-resilient computing and stochastic computing.

1.3.1 Error-Resilient Computing

The reliability of integrated circuits is a major problem in advanced semiconductor technologies. The transistors in an IC are more susceptible to transient and permanent errors at smaller feature sizes. Toward this end, multiple techniques have been proposed in the literature to ensure reliable and error-free computing—commonly referred to as error-resilient or fault-tolerant computing. For example, some techniques employ conservative designs by utilizing higher voltages and reduced operational frequencies [21]. Other techniques either utilize redundant operational units or employ error-detection and error-correction circuitries. However, these techniques trade the overall performance of a computing system for achieving an error-free computational environment. Some recent works have also focused on utilizing application-level error masking and algorithmic-level fault tolerance—instead of architectural-level techniques—to improve the overall performance of the computing systems [22].

1.3.2 Stochastic Computing

Compared to traditional computing, which processes deterministic binary strings, the stochastic computing technique processes random binary strings. Each random binary string represents the probability of $1's$ in that string [23, 24]. For example, the number 0.5 can be represented using the 8-bit string $'10101010'$. It should be noted that the representation of a number (probability) only depends on the ratio of the number of $1's$ in a string to the total number of bits in the string. Therefore, a given number can be denoted using multiple representations. The bit location insignificance also introduces fault tolerance (error resilience) in stochastic computing. However, this feature also means that an n-bit string can represent only $n + 1$ unique values, and for representing more number of unique values, higher bit width should be utilized. One of the main advantages of stochastic computing is the reduced computational complexity of the basic arithmetic operations such as addition and multiplication. For example, the multiplication of 0.5 ($'10101010'$) with 0.5 ($'11001001'$) can be achieved by performing the binary *AND* operation on the two strings to produce 0.25 ($'10001000'$). However, the binary string—generated using a stochastic number generator—has a significant impact on the accuracy of the generated results [25]. For example, performing *AND* operation on $'10101010'$ and $'10101010'$ to multiply 0.5 with 0.5 results in $'10101010'$. Similarly, performing *AND* operation on $'10101010'$ and $'01010101'$ to multiply 0.5 with 0.5 results in $'00000000'$.

1.3.3 Approximate Computing

Approximate computing is not a new concept, and it has been extensively used since the early generations of digital computers. The real-world data is analog and covers a continuous scale of values. Computers employ analog to digital converters to sample and represent the real-world data—approximate representation of data—for processing. The IEEE floating-point standard, single- and double-precision, also introduces approximations in the data representation by utilizing a fixed precision. In the recent quest for energy-efficient computing and the ubiquitous presence of error-resilient applications in every domain, approximate computing has received significant attention from both academia and industry [26]. Recent works in this direction have explored various approximation techniques—covering all layers of the computation stack—to reduce the overall computational complexity and trade output accuracy for performance gains [27, 28]. Figure 1.3 provides a broad classification of the commonly explored approximation techniques. In the following, we provide a brief description of some of the state-of-the-art works in each category. Please note that the relevant state-of-the-art work for each novel contribution of the book is provided in its respective chapter.

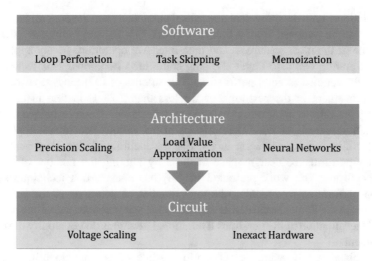

Fig. 1.3 Approximations across various layers of the computation stack

1.3.4   Software Layer Approximation

These approximations cover various techniques proposed for algorithms, programming languages, and operating systems. These techniques aim to reduce the
overall computational complexity of an application by employing different types
of heuristics and provide opportunities for efficient utilization of approximations at
other layers of the computation stack. For example, Eon presented in [29] allows the
annotation of the control flow of a program with various energy levels. Depending
upon the provided accuracy (quality) of a computational block, execution paths with
different energy levels can be opted at runtime to satisfy the overall energy constraint
of a system. The Topaz framework allows mapping the error-resilient computations
on the approximate hardware and provides a quality control mechanism to identify
the outliers in the produced output [30]. The identified outliers are re-executed on
the precise hardware to control the overall quality of the generated output. Similarly,
the EnerJ language provides an opportunity to annotate various data structures—
and the corresponding operations—as either accurate or approximate [31]. The
approximate data structures are stored and computed using approximate elements.

Among other software layer techniques, task perforation and memoization
are commonly explored techniques. Task perforation refers to the technique of
dropping—skip execution—of computationally intensive and error-resilient computations of an application. For example, the authors of [32] have provided a compiler
for identifying error-resilient codes in an application and skipping error-resilient
code segments according to the application's accuracy-performance constraints. In
recent years, many works have targeted task perforation for reducing the computational complexity and storage requirements of Deep Neural Networks (DNNs). In
the context of DNNs, this technique is commonly referred to as *network pruning*. In

this technique, both the individual noncritical parameters and the noncritical neurons (filters) can be removed from a trained network [33–36]. As *loops executions* are often one of the most resources- and time-intensive operations in any application, task perforation can also target skipping some loop iterations. This technique is commonly denoted as loop perforation. The authors of [37] have reported $2\times$ to $7\times$ improvement in the performance for less than 10% reduction in the output accuracy for various applications using loop perforations. Their technique performs a criticality analysis of the different loops in the various benchmark applications and identifies the error-tolerant loops. For the identified loops, they perform an accuracy-performance design space analysis by altering the number of dropped loop iterations. The work presented in [38] has also used a technique similar to loop perforation by reducing the total number of iterations of the refinement algorithm in the K-means clustering application. Their proposed technique performs an early termination of the iterations if additional iterations do not modify the cluster assignment.

Memoization is the process of storing results of computationally expensive operations and utilizing these results later for identical operations and input data. The approximate memoization technique can be used to produce approximate results for new input data, which show some correlation with the data for which a result has already been computed and stored. For example, the authors of [39] have exploited the spatial locality of data values in single-instruction-multiple-data (SIMD) architectures to propose a spatial memoization technique. Their proposed technique performs an error-free operation on an instance of data, and the output value of the operation is used to correct variation-induced timing errors on other occurrences of the same (or spatially located) data. The work presented in [40] has also utilized memoization to provide results of operations utilizing approximately similar input data for graphics applications.

1.3.5 Architecture Layer Approximation

Modern computing systems—from high-computing servers to embedded processors—utilize 32-bit and 64-bit operations. These systems utilize various number representation schemes such as IEEE single- and double-precision floating point and 32-bit and 64-bit integer schemes to perform these operations. However, such high-precision number schemes are often not required for most error-resilient applications, such as image processing and machine learning models. Moreover, these numbering schemes' corresponding computational complexity and storage requirements are also serious limitations for resource-constrained embedded systems. Toward this end, the utilization of precision scaling is one of the most explored approximation techniques. For example, the BFloat16 (16-bit floating-point numbers) developed by Google Brain is utilized in many modern systems and frameworks such as Google cloud TPUs, Intel Deep Learning Boost, and TensorFlow [41–43]. Many recent works have shown the exploration of low precision numbers (8-bit and below) for reducing the computational complexity of ANNs [20, 44, 45]. Toward this end, many recent works have proposed various quantization schemes to efficiently utilize the available bit width [14, 46].

Modern processors—multi-core and heterogeneous architectures—utilize multiple levels of caches to satisfy the data requirements of various computations. These caches are responsible for alleviating the long latency and high-energy consumption of accessing data from the main memory. However, despite utilizing multiple levels of cache and a large amount of last level cache, a cache miss (data not available in cache) is a frequently occurring phenomenon in memory load operations. Toward this end, many recent works have proposed load value approximation techniques to avoid the long latency and energy consumption incurred in fetching data from the main memory for error-resilient applications. In this technique, an approximate value is generated and provided to the computational units in case of a cache miss. For example, the authors of [47] have presented an approximate and rollback-free load value approximator for error-resilient applications. Their proposed implementation provides two accuracy-performance knobs to control the accuracy of load value approximator and provide different trade-offs between output accuracy and performance of an application. The first performance knob, *the confidence window*, defines the maximum allowed difference between actual and approximate values. The second knob, *the approximation degree*, defines the number of times the approximate value is generated and utilized before fetching the actual value from memory and utilizing it to train the approximator. The Doppelgänger cache proposed in [48] utilizes values similarity index across different data blocks to identify data blocks having similar data. For all the identified blocks, a single data block is stored in the cache, and the associated tags of the identified data blocks are mapped to the single stored entry in the cache. Therefore, all subsequent requests to data in any of the identified blocks are mapped to one single stored cache block. The Doppelgänger cache also utilizes an accurate cache to accommodate the accuracy-sensitive data, and the approximate cache is utilized to store only the error-resilient data. The authors of [49] have also proposed a rollback-free value prediction technique to propose an approximate load value predictor for Graphics Processing Units (GPUs). Their proposed technique utilizes data similarity feature between spatially correlated threads in a SIMD architecture. Their proposed technique provides a *Drop Rate* performance knob to control the number of cache misses that utilize the value provided by the predictor.

ANNs are capable of representing the functionality of a wide range of functions [50]. Considering the high parallelism offered by ANNs, they can be used to approximate the error-resilient functions of an application. However, in many cases, the invocation of an ANN code (to replace an error-resilient function in an application) may degrade the application's overall performance. For example, the authors of [51] analyzed 11 different error-resilient benchmark applications and identified that these applications spend on average 56% of their runtime and 59% of their energy in functions that ANNs can replace. However, the invocation of ANNs for these functions on the same hardware platform results in $3.2\times$ degraded performance. Therefore, it is essential to design efficient Neural Processing Units

(NPUs) for the acceleration of ANNs. For example, the authors of [52] have proposed a NPU to provide ANN accelerations for general-purpose processing systems. Their NPU can implement various Multi-layer Perceptrons (MLPs) as decided by the accuracy and performance requirement of the application. The work presented in [51] describes an MLP-based NPU for GPUs. Similarly, the authors of [53] have used FPGAs to design an MLP-based NPU for accelerating the error-tolerant functions.

1.3.6 Circuit Layer Approximation

Circuit-level approximations exploit various techniques to implement inaccurate (approximate) memory modules and computational nodes. From a cross-layer approximation perspective, the utilization of approximate circuits along with approximations at other layers of the computation stack can significantly improve the overall performance of an implementation [9, 38]. In this section, we summarize the commonly used techniques to design approximate circuits.

Dynamic Voltage and Frequency Scaling (DVFS) is a commonly explored technique to reduce the energy consumption of a circuit. For a digital circuit, dynamic power consumption P_d is defined as [28]

$$P_d \ \propto \ CV_d^2 f \tag{1.1}$$

where V_d is the supply voltage, C is the effective capacitance of the circuit, and f shows the operational frequency of the circuit. Given the quadratic relation between V_d and P_d, the dynamic power consumption of a circuit can be significantly reduced by decreasing the supply voltage. This technique is commonly referred to as *voltage overscaling*. However, reducing the supply voltage also increases the time to charge the effective load capacitance of the circuit, which affects the maximum allowed operating frequency of the circuit. DVFS is a commonly employed technique in many modern architectures to match the dynamic power consumption of a system with the desired performance [54–56]. For a given operational frequency f, voltage overscaling can result in the nondeterministic timing violations of a computational circuit and bit errors in memory units. However, many recent works have exploited the error resilience of application to use voltage overscaling for reducing the overall energy consumptions of different applications. For example, the authors of [57] have proposed partially forgetful cache memories by segmenting the cache memory into reliable and nonreliable parts. The reliable segment of the memory is operated using suitable voltage, and the nonreliable portion of the memory is operated using a lower voltage. The nonreliable memory segment (approximate) is then utilized for storing error-resilient data structures of an application. The authors of [58] have used voltage overscaling to design an approximate associative memristive memory for storing the results of frequently used operations in a GPU-based system. The work presented in [59] utilizes an adaptive voltage overscaling technique to dynamically switch a texture decompression accelerator's execution mode between accurate and approximate. In this technique, the voltage is reduced, and an error counter is used to accumulate the generated errors. Based on the provided error threshold, the voltage is raised again. Another example of utilizing voltage overscaling for energy efficiency is the recognition and mining processor presented in [38]. It utilizes voltage overscaling for designing approximate Multiply-Accumulate (MAC) units. However, as noted by [28, 60], the errors generated by voltage overscaling are nondeterministic and extra resources are required to control the accuracy of the computations.

Among other circuit-level approximation techniques, circuit pruning is a commonly explored method. Circuit pruning refers to the technique of removing or simplifying (approximating) parts of a circuit based on their significance. As described in [6], the focus of approximations, in this case pruning, should be the components (logic) that are among the major contributors to an implementation's overall performance metrics—utilized resources, critical path delay, and power consumption—and have little impact on the output accuracy of the circuit. As highlighted in [60], arithmetic circuits are a good candidate for such types of approximations because of their bit-significance notion. The frequent utilization of the arithmetic operations, such as addition and multiplication, in various error-resilient applications also signifies the need for their approximations. For example, the VGG-16 DNN utilizes $15.5G$ MAC operations to process one single 224×224 input image [61]. Toward this end, many recent works have proposed various approximate designs for different arithmetic operators. These approximate designs are then used in the implementation of approximate hardware accelerators for various applications. Here we provide a summary of some of the related approximate arithmetic circuits.

The authors of [10] have utilized transistor pruning to propose approximate full adders with fewer transistors, reduced critical path delay, and power consumption. Their technique judiciously removes transistors from an accurate adder to propose five different approximate adders. These adders are then utilized to implement multi-bit adders for image and video compression algorithms. The authors of [62] have also used transistor-level pruning to implement three approximate adders from two accurate adder designs. The work presented in [60] has proposed a framework for gate-level pruning of arithmetic operators. The framework analyzes the generated netlist of an accurate implementation of an operator and ranks the different gates based on their significance and switching activity (toggle count). The framework assigns location-based significance to the gates in the netlist. For this purpose, it assigns two times higher significance to all the gates contributing to the computation of output bit O_{i+1} than those contributing to the generation of output bit O_i. The toggle count is computed by simulating the design and extracting the corresponding switching rate for each node. While implementing approximate operators, the gates with lower rankings are pruned first. The authors have evaluated the generated designs' efficacy by implementing them in an accelerator for discrete cosine transform operation. Many approximate adder designs have considered eliminating long carry propagation paths to reduce the overall critical path delay of the design.

For example, the approximate adder GeAr presented in [63] utilizes multiple sub-adders to implement a larger adder. All sub-adders operate in parallel, and there is no carry propagation from one sub-adder to the other. However, except for the least significant adder, all adders utilize a number of previous bits to estimate the missing carry. For this purpose, GeAr adder utilizes the number of bits to predict missing carry as a design-time parameter. GeAr also supports an error correction circuitry to produce accurate output if required. However, the error correction circuitry decreases the overall throughput of the adder. The authors of [64] and [65] have used various Lookup Table (LUT)-level approximations to propose approximate adders optimized for FPGA-based systems. These adders also utilize the carry propagation elimination to reduce the adders' critical path delay and propose various techniques to estimate the missing carries.

Many works in the domain of circuit-level approximations have also focused on designing approximate multipliers. A multiplication operation consists of generating partial products according to some defined algorithm and then adding the partial products to compute the final product. For example, Eq. 1.2 defines an algorithm for implementing M $\times$ N unsigned multiplication of two integers. This multiplication algorithm generates all the partial products by performing logical *AND* operations on all the bits of both operands. The generated partial products can be added using various techniques to compute the final product. For example, Fig. 1.4 presents an example of a 2 $\times$ 2 carry-save adder-based array multiplier [66]. In this technique, each computing unit, represented by the square boxes in the figure, computes a partial product term and adds it with the sum and previous signals from a computing unit in the previous row. The generated partial products can also be added using a tree of compressors to implement a tree-based multiplier. In this regard, Wallace and Dadda trees are commonly used techniques to compute the final products [67, 68]. Figure 1.5 shows the first stages of the Wallace and Dadda compression tree to reduce the four rows of partial products (for a 4 $\times$ 4 multiplier) to three rows. The rectangular boxes in the figure show the utilization of half and full adders. Here we provide an overview of some of the recent works that have focused on the various sub-operations of multiplications to design approximate multipliers.

$$A \times B = \sum_{m=0}^{M-1} \sum_{n=0}^{N-1} a_n b_m 2^{m+n} \tag{1.2}$$

The authors of [69] have reduced the computational complexity of multiplier by perforating the generation of multiple consecutive partial product rows. The proposed design utilizes two design-time parameters, which identify the index of the first perforated row and the numbers of rows to be perforated, respectively. A similar work, proposed in [70], truncates the partial product terms in a carry-save adder-based implementation. This work identifies the partial products in an M $\times$ N multiplier by utilizing horizontal and vertical indices. All the partial product terms positioned either above the horizontal index or to the right of the vertical index are omitted to compute the approximate product. The M $\times$ N approximate

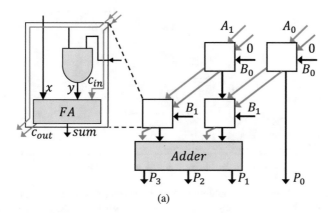

Fig. 1.4 Carry-save adder-based unsigned 2×2 multiplier [66]

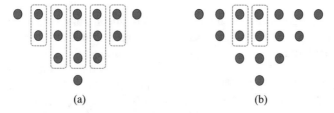

(a)   (b)

Fig. 1.5 First stage of reduction of Wallace and Dadda trees. (**a**) Wallace tree. (**b**) Dadda tree

multiplier proposed in [71] separates the input operands into nonsignificant and significant parts. The significant parts of the operands (the most significant bits) are computed using a K × L accurate multiplier where K ≤ M and L ≤ N. The nonsignificant part of the product is computed without actual multiplication. For this purpose, the nonsignificant parts of the operands are analyzed from left to right—higher significance to lower significance—to find the leading 1 location in any of the operands. All product bits corresponding to the leading 1 location and the subsequent locations are assigned constant $1's$. The remaining nonsignificant product bits are assigned constant $0's$. The work presented in [72] also utilizes a smaller K × K accurate multiplier to implement an approximate N × N multiplier where $N > K$. The proposed implementation utilizes leading 1 detectors to dynamically identify and extract the most significant K-bit values of the two operands. The extracted K-bits are provided to the accurate K × K multiplier to produce a $2K$-bit output, and the remaining product bits are computed approximately. The works presented in [73] and [74] have also utilized leading 1 detectors to propose unsigned approximate multipliers. These works have utilized approximate $\log_2$ to utilize addition operations for computing the approximate product.

The authors of [75] have utilized the modular design approach to implement approximate multipliers. In this technique, a larger multiplier is implemented by

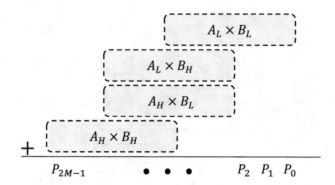

Fig. 1.6 Designing higher-order multipliers from lower-order multipliers: A_L and A_H represent $\frac{M}{2}$-bit LSBs and MSBs, respectively, of M-bit A. B_L and B_H represent $\frac{M}{2}$-bit LSBs and MSBs, respectively, of M-bit B

accumulating the products of multiple sub-multipliers. For example, as shown in Fig. 1.6, the results of four $\frac{M}{2} \times \frac{M}{2}$ sub-multipliers are added together to implement an M × M multiplier. The sub-multiplier $A_H \times B_L$ utilizes the MSBs of the input A; therefore, the result of the $A_H \times B_L$ should be shifted left by $\frac{M}{2}$-bits before addition with the result of $A_L \times B_L$. Similarly, the multiplier $A_L \times B_H$ utilizes the MSBs of operand B, and its result should also be shifted by $\frac{M}{2}$-bits before addition. The multiplier $A_H \times B_H$ utilizes the MSBs of both operands; therefore, its generated result should be shifted left by M-bits before addition. The approximate multiplier design presented in [75] utilizes an accurate multiplier for $A_H \times B_H$ to avoid significant degradation in the output accuracy. The remaining three sub-multipliers are approximate designs. Toward this end, the authors have utilized approximate Wallace tree reduction to compute the approximate products of the sub-multipliers. The modular design approach is also utilized by some other works such as [11] and [76], which propose approximate 2 × 2 multipliers to implement higher-order multipliers.

In a tree-based multiplier, the reduction of the partial products is the most computational-intensive process. The Wallace and Dadda tree techniques, as shown in Fig. 1.5, utilize 3:2 compressors (full adders) and 2:2 compressors (half adders) to reduce the generated partial products. Some recent works have focused on designing various approximate compressors to improve the overall performance of the reduction stage. The approximate 4:2 compressor is one of the most explored techniques in this regard. An accurate 4:2 compressor generates 2-bit output (sum and carry) and an output C_{out} [77]. Only when all input bits are 1's, a 4:2 compressor utilizes all output bits and generates an output of 100_2. The approximate 4:2 compressor proposed in [78] approximates the 100_2 output with 10_2 to reduce the computational complexity of the compressor. To reduce the output error of the approximate 4:2 compressors, the authors of [77] and [79] compute the propagate and generate signals from the generated partial product terms. The approximate compressors then utilize these signals to compute the approximate product.

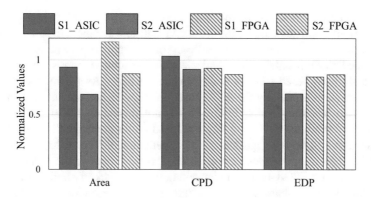

Fig. 1.7 Cross-platform comparison of area, latency, and EDP of 8×8 unsigned multipliers: results are normalized to the corresponding results of ASIC- and FPGA-based accurate multipliers

1.4   Problem Statement

The short time-to-market, low non-recurring engineering cost, runtime reconfig-urability, and high parallelism offered by FPGAs make them an attractive choice for implementing hardware accelerators. It allows the utilization of the same silicon chip to implement different functions by reconfiguring it. Today FPGAs are increasingly used in computing systems ranging from embedded devices at the edge to data centers and cloud servers. For example, Xilinx provides an edge AI platform to implement various machine learning models on embedded devices at the edge [80]. Similarly, companies like Microsoft, Amazon, Intel, Alibaba, and Huawei utilize FPGAs in their data centers and cloud servers [81–85]. The current generations of FPGAs provide dedicated functional blocks, such as Digital Signal Processing (DSP) blocks, to offer ASIC-like performance and energy efficiency for various operations. However, despite all the technological and architectural innovations, FPGA-based implementations consume more power than ASIC-based implementations. Toward this end, the paradigm of approximate computing appears a feasible solution to improve the overall resource utilization, latency, and energy consumption of FPGA-based hardware accelerators for error-resilient applications.

Many recent works, such as [11, 72, 76, 86], have proposed various approximate arithmetic units for the hardware implementation of error-resilient applications. However, most of these works consider only ASIC-based systems. Due to the inherent architectural differences between FPGAs and ASICs, these ASIC-based designs provide limited or no performance gains when directly synthesized for the FPGA-based systems. For example, Fig. 1.7 compares the ASIC-based area, Critical Path Delay (CPD), and energy-delay-product (EDP) of two state-of-the-art approximate multipliers, *"S1"*, presented in [76], and *"S2"*, described in [11], with their FPGA-based implementations. Both of these designs have been originally presented for ASIC-based systems. The ASIC-based implementation results have been obtained from the corresponding papers [11] and [76], whereas for the FPGA-based implementations, the Xilinx Vivado tool for the Virtex-7 family has been used. Further, to evaluate the efficacy of the approximate designs, we have normalized these results to the implementation results of corresponding ASIC-based and FPGA-based accurate multipliers, respectively. As shown by the analysis results, the gains offered by the ASIC-based implementation are not proportionally translated to the corresponding FPGA-based implementation. For example, the area and EDP gains offered by $S1$ and $S2$ are reduced for the corresponding FPGA-based implementations; in fact, approximate implementation of $S1$ consumes more FPGA resources than the corresponding accurate design. However, the CPD is further reduced for both FPGA-based implementations. This lack of similar performance gains for the FPGA-based systems is the result of the architectural differences between ASICs and FPGAs. In ASIC-based designs, logic gates are deployed for the implementations of different logic circuits; thus, a full control over resource utilization at a fine granularity is possible. However, FPGA-based computational blocks are composed of entirely different entities, i.e., LUTs, where configuration bits are used to implement an individual circuit.

1.4.1 Research Challenge

Considering the high utilization of FPGAs across the computation spectrum and, in particular, for providing acceleration for error-resilient applications, this book focuses on the following research challenge:

> *Design and analysis of LUT-level optimizations to define accurate and approximate arithmetic blocks for FPGA-based accelerators for error-resilient applications.*

Since multiplication is one of the widely used arithmetic operations in various applications, such as image/video processing and machine learning, this book has mainly focused on the accurate and approximate designs of multipliers. Further, we also present a methodology to design FPGA-optimized approximate adders. We further divide the research problem abovementioned into the following research questions:

- FPGA synthesis tools provide high-performance DSP blocks for multiplication and MAC operations. Therefore, is there a need to design LUT-based soft multipliers and adders? (Addressed in Chap. 3)
- FPGA synthesis tools also provide various area- and speed-optimized soft multiplier Intellectual Properties (IPs). Can the nature of the multiplication—unsigned, signed, and constant—be utilized to implement multiplier IPs having better performance (resource utilization, latency, and energy consumption) than those offered by vendor IPs? (addressed in Chaps. 3 and 4)

- FPGA synthesis tools utilize LUTs and associated carry chains for implementing all types of combinational circuits. What factors should be considered while utilizing these primitives for designing an accurate/approximate operator to improve the overall performance of the implementation? (addressed in Chaps. 3 and 4)
- Different error-resilient applications have different tolerance to the various approximations. Therefore, a single approximate operator may not satisfy the required accuracy-performance constraints when used in an error-resilient application. Toward this end, can a generic design methodology be implemented to provide approximate operators according to an application's inherited error resilience and required accuracy-performance constraints? (addressed in Chap. 5)
- An operator's performance is characterized by various parameters such as resource utilization, critical path delay, and power consumption. These parameters can be utilized to estimate the performance impact of deploying an operator in the implementation of an application. Similarly, the accuracy of an approximate operator is typically characterized by various statistical error metrics such as the total number of error occurrences, the maximum error magnitude, and the average error magnitude. However, the utilization of these error metrics to estimate the accuracy impact of deploying an operator in an application is mostly unknown. Further, the actual deployment of an approximate operator, from a library of approximate operators, in an application to perform the accuracy analysis is often a time- and resource-consuming operation. Toward this end, can a methodology be designed to evaluate the impact of various approximate operators on an application's output accuracy? (addressed in Chap. 6)

Besides mainly focusing on operator-level approximations to exploit error resilience of various applications, this book also explores other layers of the computation stack for possible approximations. In particular, the following aspect is explored:

- The accuracy-performance constraints of an application may not be satisfied (or exhaustively explored) by utilizing only circuit-level approximations. To this end, can a cross-layer approximation analysis framework be designed to provide fast estimates of the behavioral and performance impact of various approximations on the different layers of the computation stack? (addressed in Chap. 6)

1.5 Focus of the Book

Figure 1.8 presents the overall structure of the book to address the research challenges mentioned above. We utilize the FPGA synthesis tool provided DSP blocks- and LUTs-based multipliers for two different applications to motivate the need for soft logic-based arithmetic operators. Based on our motivational analysis, we present various designs of accurate multipliers in Chap. 3. Our presented designs utilize 6-input LUTs and associated carry chains (explained in Chap. 2) for resource-

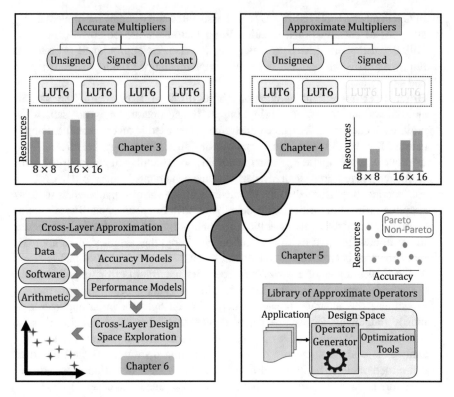

Fig. 1.8 Overall structure of the book

efficient, low-latency, and energy-efficient implementations. As an application may utilize *unsigned*, *signed*, or *constant* numbers-based multipliers, we explore various multiplication algorithms (presented in Chap. 2) and provide optimized LUT-based implementations of *unsigned*, *signed*, and *constant* multipliers. Our proposed designs are scalable and can be used to implement accurate multipliers of arbitrary size. We have characterized and compared our presented designs with state-of-the-art designs using various standard performance metrics, such as LUT utilization, Critical Path Delay (CPD), and power/energy consumption. We further evaluate the efficacy of the multipliers by implementing them in various applications, such as ANNs and Reinforcement Learning (RL).

The proposed accurate *unsigned* and *signed* designs are analyzed in Chap. 4 to propose various approximate multiplier designs. The proposed approximations target the reduction in total utilized LUTs, CPD, and power/energy consumption of the multipliers by reducing the accuracy of the multiplier. To characterize the output accuracy of the presented approximate multipliers, we have used commonly used statistical error metrics, such as average relative error value and average error magnitude. The various error metrics utilized in this book are described in Chap. 2. We provide an exhaustive accuracy-performance analysis of the proposed designs with state-of-the-art approximate multipliers. Further, we have also utilized applications from various domains, such as image processing and machine learning, to evaluate the impact of approximations on the output accuracy of the application and the performance of the corresponding implementation.

As mentioned in the research questions, an application-agnostic design methodology of implementing approximate operators may result in approximate operators that may not satisfy the specified accuracy-performance constraints of specific applications. Therefore, it is necessary to design a framework that can analyze and exploit the error resilience of applications to provide some information about the approximate operators capable of satisfying the application-level accuracy-performance constraints. Toward this end, there is also the need for a generic design methodology to implement approximate operators based on the provided information. We address these challenges in Chap. 5 by utilizing Multi-objective Bayesian Optimization (MBO) and Genetic Algorithm (GA) (described in Chap. 2) to implement a framework that performs an application-level accuracy-performance analysis for identifying feasible approximate operators. Our framework utilizes a *binary string* to represent the identified approximate operators. For example, in the case of approximate multipliers, the *binary string* represents the multiplier configurations that can satisfy accuracy-performance constraints for an application. We further develop a generic design methodology for implementing LUTs-based approximate operators according to provided configurations. Compared to most state-of-the-art works, such as [87], that perform a design space exploration using an existing library of approximate operators for identifying feasible approximate operators for an application, our proposed methodology implements new approximate operators for the application under consideration. The proposed framework and operator generation methodology are generic and can be used for any application and arithmetic operator. In this book, we have used it to implement approximate adders and multipliers.

Finally, Chap. 6 in the book focuses on providing a framework for an efficient analysis of the behavioral impact of utilizing available libraries of approximate operators in error-resilient applications. One of the possible solutions in this regard is the actual deployment of behavioral models of approximate operators in a high-level implementation of an application and then the evaluation of the impact using various testbench datasets. However, the actual behavioral execution of an application for a large set of approximate operators is a time- and resource-consuming operation. Toward this end, a feasible solution can be utilizing various trained Machine Learning (ML) models to evaluate the behavioral impact. However, identifying and selecting parameters that can represent an approximate operator in an ML model is mostly unknown. Toward this end, we present a novel Polynomial Regression (PR)-based technique for modeling approximate operators. Utilizing the PR coefficients, we can represent various approximate arithmetic operators with high accuracy. In this book, we have used it to represent approximate adders and multipliers. We utilize these coefficients to train ML models to evaluate approximate operators' impact on an application's output accuracy. This book utilizes Gaussian image smoothing as a test application for analyzing approximate multipliers. Our analysis shows that PR coefficients-based ML models can also estimate, with high accuracy, the application-level impact of approximate multipliers that are not present in the training dataset of ML models. The proposed framework also allows integrating approximations at other layers of the computation stack to enable a *cross-layer approximation analysis framework*. For example, for the test application, the framework also analyzes the behavioral impact of algorithmic and software-level approximations such as mode of convolution ($1D/2D$), the stride length of the convolution, and convolution kernel window size. Further, the framework provides an Multi-objective Bayesian Optimization (MBO)-based Design Space Exploration (DSE) methodology to identify feasible design points that offer better trade-offs between application error and hardware performance.

1.6 Key Contributions and Book Overview

The major contributions of the book are listed below. These contributions are open source and available at https://cfaed.tu-dresden.de/pd-downloads to facilitate reproducible results and fuel further research in this direction.

- *FPGA-optimized softcore accurate multipliers (Chap. 3):* The book presents various designs of unsigned, signed, and constant multipliers. These designs are based on the work published in *IEEE Transactions on Computer-Aided Design of Integrated Circuits and Systems (TCAD)-2021* [88], *IEEE Transactions on Computers (TC)-2021* [89], *IEEE Embedded Systems Letters-2021* [90], and *IEEE TCAD-2021* [91].
- *FPGA-optimized softcore approximate multipliers (Chap. 4):* We present various designs of approximate unsigned and signed multipliers. These designs are based on the work published in the *Proceedings of the Annual Design Automation Conference (DAC)-2018* [92], *Proceedings of DAC-2018* [93], *IEEE TCAD-2021* [88], and *IEEE TC-2021* [89]. Ullah et al. [88] is an extended version of [92].
- *Application-specific approximate operators (Chap. 5):* We present a generic methodology for the implementation and design space exploration of application-specific approximate operators, such as adders and multipliers. This contribution of the book is published in *ACM Transactions on Embedded Computing Systems* [94].
- *A highly accurate statistical representation of approximate operators (Chap. 6):* The book presents a polynomial regression-based technique to model approximate operators. This work is published in *Proceedings of DAC-2021* [95].
- *Analysis framework for cross-layer approximation (Chap. 6):* We utilize various machine learning models to analyze the behavioral and performance impact of approximations at various layers of the computation stack. This contribution is based on the work published in the *Proceedings of DAC-2021* [95].

The rest of the book is organized as follows. Chapter 2 provides the required preliminary information to understand better the various contributions presented in other chapters. Specifically, it provides an overview of the slice structure of Xilinx FPGAs, introduces different signed multiplication algorithms, and defines the various statistical error metrics utilized in this book to characterize approximate operators, and two optimization algorithms (Genetic Algorithm (GA) and MBO) were employed for design space exploration of approximate circuits. Chapter 3 presents various designs of accurate unsigned, signed, and constant coefficient-based multipliers. These designs are based on the efficient utilization of 6-input LUTs and associated carry chains of the FPGAs. These designs are further analyzed in Chap. 4 to present various approximate multiplier designs that trade the multiplication output accuracy to improve the overall performance of the implementations. Chapter 5 provides a generic methodology for implementing approximate operators. It also presents a framework that utilizes various machine learning models for implementing application-specific approximate operators. Chapter 6 describes our polynomial regression-based representation of approximate operators and utilizes it to present a cross-layer approximation analysis framework. Finally, Chap. 7 concludes the book and provides suggestions about future work in this domain.

References

1. G.E. Moore et al., *Cramming More Components Onto Integrated Circuits* (McGraw-Hill, New York, 1965).
2. M. Bohr, The new era of scaling in an SoC world, in *2009 IEEE International Solid-State Circuits Conference—Digest of Technical Papers* (2009), pp. 23–28
3. R.H. Dennard, F.H. Gaensslen, H.-N. Yu, V.L. Rideovt, E. Bassous, A.R. Leblanc, Design of ion-implanted MOSFET's with very small physical dimensions. IEEE Solid-State Circuits Soc. Newslett. **12**(1), 38–50 (2007)
4. C. Martin, Multicore processors: challenges, opportunities, emerging trends, in *Proc. Embedded World Conference*, vol. 2014 (2014), p. 1
5. H. Esmaeilzadeh, E. Blem, R. St. Amant, K. Sankaralingam, D. Burger, Dark silicon and the end of multicore scaling, in *2011 38th Annual International Symposium on Computer Architecture (ISCA)* (IEEE, Piscataway, 2011), pp. 365–376
6. S. Venkataramani, S.T. Chakradhar, K. Roy, A. Raghunathan, Computing approximately, and efficiently, in *2015 Design, Automation & Test in Europe Conference & Exhibition (DATE)* (IEEE, Piscataway, 2015), pp. 748–751
7. Apple A14. https://en.wikipedia.org/wiki/Apple_A14. Accessed 20 Aug 2021
8. V.K. Chippa, S.T. Chakradhar, K. Roy, A. Raghunathan, Analysis and characterization of inherent application resilience for approximate computing, in *2013 50th ACM/EDAC/IEEE Design Automation Conference (DAC)* (2013), pp. 1–9
9. M. Shafique, R. Hafiz, S. Rehman, W. El-Harouni, J. Henkel, Invited: cross-layer approximate computing: from logic to architectures, in *2016 53nd ACM/EDAC/IEEE Design Automation Conference (DAC)* (2016), pp. 1–6
10. V. Gupta, D. Mohapatra, A. Raghunathan, K. Roy, Low-power digital signal processing using approximate adders. IEEE Trans. Comput.-Aided Design Integr. Circuits Syst. **32**(1), 124–137 (2013)

11. P. Kulkarni, P. Gupta, M. Ercegovac, Trading accuracy for power with an underdesigned multiplier architecture, in *2011 24th Internatioal Conference on VLSI Design* (IEEE, Piscataway, 2011), pp. 346–351

12. V. Mrazek, R. Hrbacek, Z. Vasicek, L. Sekanina, EvoApprox8b: library of approximate adders and multipliers for circuit design and benchmarking of approximation methods, in *Design, Automation Test in Europe Conference Exhibition (DATE), 2017* (2017), pp. 258–261

13. V. Mrazek, L. Sekanina, Z. Vasicek, Libraries of approximate circuits: automated design and application in CNN accelerators. IEEE J. Emerg. Sel. Topics Circuits Syst. **10**(4), 406–418 (2020)

14. S. Vogel, J. Springer, A. Guntoro, G. Ascheid, Selfsupervised quantization of pre-trained neural networks for multiplierless acceleration, in *2019 Design, Automation & Test in Europe Conference & Exhibition (DATE)* (IEEE, Piscataway, 2019), pp. 1094–1099

15. The number glitch that can lead to catastrophe. https://www.bbc.com/future/article/20150505-the-numbers-that-lead-to-disaster. Accessed 1 Jan 2021

16. A.K. Mishra, R. Barik, S. Paul, iACT: a softwarehardware framework for understanding the scope of approximate computing, in *Workshop on Approximate Computing Across the System Stack (WACAS)* (2014), p. 52

17. P. Roy, R. Ray, C. Wang, W.F. Wong, Asac: automatic sensitivity analysis for approximate computing, in *Proceedings of the 2014 SIGPLAN/SIGBED Conference on Languages, Compilers and Tools for Embedded Systems* (2014), pp. 95–104

18. A. Yazdanbakhsh, D. Mahajan, H. Esmaeilzadeh, P. Lotfi-Kamran, AxBench: a multiplatform benchmark suite for approximate computing. IEEE Design Test **34**(2), 60–68 (2016)

19. S. Zhou, Y. Wu, Z. Ni, X. Zhou, H. Wen, Y. Zou, Dorefa-net: training low bitwidth convolutional neural networks with low bitwidth gradients (2016). arXiv preprint arXiv:1606.06160

20. M. Courbariaux, Y. Bengio, J.-P. David, Binaryconnect: training deep neural networks with binary weights during propagations, in *Advances in Neural Information Processing Systems* (2015), pp. 3123–3131

21. S. Das, D.M. Bull, P.N. Whatmough, Error-resilient design techniques for reliable and dependable computing. IEEE Trans. Device Mater. Reliab. **15**(1), 24–34 (2015)

22. L. Leem, H. Cho, J. Bau, Q.A. Jacobson, S. Mitra, ERSA: error resilient system architecture for probabilistic applications, in *2010 Design, Automation Test in Europe Conference Exhibition (DATE 2010)* (2010), pp. 1560–1565

23. F. Neugebauer, I. Polian, J.P Hayes, On the limits of stochastic computing, in *2019 IEEE International Conference on Rebooting Computing (ICRC)* (IEEE, Piscataway, 2019), pp. 1–8

24. E. Vahapoglu, M. Altun, *From Stochastic to Bit Stream Computing: Accurate Implementation of Arithmetic Circuits and Applications in Neural Networks* (2019). arXiv: 1805.06262 [cs.ET]

25. S.A. Salehi, Low-cost stochastic number generators for stochastic computing. IEEE Trans. Very Large Scale Integr. Syst. **28**(4), 992–1001 (2020)

26. W. Liu, F. Lombardi, M. Shulte, A retrospective and prospective view of approximate computing [point of view]. Proc. IEEE **108**(3), 394–399 (2020)

27. S. Mittal, A survey of techniques for approximate computing. ACM Comput. Surv. **48**(4), 1–33 (2016)

28. P. Stanley-Marbell et al., Exploiting errors for efficiency: a survey from circuits to applications. ACM Comput. Surv **53**(3), 1–39 (2020)

29. J. Sorber, A. Kostadinov, M. Garber, M. Brennan, M.D. Corner, E.D. Berger, Eon: a language and runtime system for perpetual systems. in *Proceedings of the 5th International Conference on Embedded Networked Sensor Systems*. SenSys '07 (Association for Computing Machinery, Sydney, 2007), pp. 161–174

30. S. Achour, M.C. Rinard, Approximate computation with outlier detection in topaz. ACM SIGPLAN Not. **50**(10), 711–730 (2015)

31. A. Sampson, W. Dietl, E. Fortuna, D. Gnanapragasam, L. Ceze, D. Grossman, EnerJ: approximate data types for safe and general low-power computation. ACM SIGPLAN Not. **46**(6), 164–174 (2011)

32. H. Hoffmann, S. Misailovic, S. Sidiroglou, A. Agarwal, M. Rinard, Using code perforation to improve performance, reduce energy consumption, and respond to failures (2009)
33. F. Manessi, A. Rozza, S. Bianco, P. Napoletano, R. Schettini, Automated pruning for deep neural network compression, in *2018 24th International Conference on Pattern Recognition (ICPR)* (2018), pp. 657–664
34. Y. He, X. Zhang, J. Sun, Channel pruning for accelerating very deep neural networks (2017). arXiv: 1707.06168 [cs.CV]
35. H. Li, A. Kadav, I. Durdanovic, H. Samet, H.P. Graf, Pruning Filters for Efficient ConvNets (2017). arXiv: 1608.08710 [cs.CV]
36. Y. Lin, Y. Tu, Z. Dou, An improved neural network pruning technology for automatic modulation classification in edge devices. IEEE Trans. Veh. Technol. **69**(5), 5703–5706 (2020)
37. S. Sidiroglou-Douskos, S. Misailovic, H. Hoffmann, M. Rinard, Managing performance vs. accuracy trade-offs with loop perforation, in *Proceedings of the 19th ACM SIGSOFT symposium and the 13th European conference on Foundations of software engineering* (2011), pp. 124–134
38. V.K. Chippa, D. Mohapatra, K. Roy, S.T. Chakradhar, A. Raghunathan, Scalable effort hardware design. IEEE Trans. Very Large Scale Integr. Syst. **22**(9), 2004–2016 (2014)
39. A. Rahimi, L. Benini, R.K. Gupta, Spatial memoization: concurrent instruction reuse to correct timing errors in SIMD architectures. IEEE Trans. Circuits Syst. II: Exp. Briefs **60**(12), 847–851 (2013)
40. G. Keramidas, C. Kokkala, I. Stamoulis, Clumsy value cache: an approximate memoization technique for mobile GPU fragment shaders, in *Workshop on Approximate Computing (WAPCO'15)* (2015)
41. S. Wang, P. Kanwar, BFloat16: the secret to high performance on Cloud TPUs, in *Google Cloud Blog* (2019)
42. Intel, *bfloat16—Hardware Numerics Definition* (2018). https://software.intel.com/content/www/us/en/develop/download/bfloat16-hardwarenumerics-definition.html. Accessed 6 Sept 2021
43. TensorFlow. *Mixed Precision* (2021). https://www.tensorflow.org/guide/mixedprecision. Accessed 6 Sept 2021
44. S. Yin, G. Srivastava, S.K. Venkataramanaiah, C. Chakrabarti, V. Berisha, J.S. Seo, Minimizing area and energy of deep learning hardware design using collective low precision and structured compression (2018). arXiv: 1804.07370 [cs.NE]
45. B. Moons, K. Goetschalckx, N. Van Berckelaer, M. Verhelst, Minimum energy quantized neural networks (2017). arXiv: 1711.00215 [cs.NE]
46. D. Lin, S. Talathi, S. Annapureddy, Fixed point quantization of deep convolutional networks, in *International Conference on Machine Learning* (PMLR, 2016), pp. 2849–2858
47. J.S. Miguel, M. Badr, N.E. Jerger, Load value approximation, in *2014 47th Annual IEEE/ACM International Symposium on Microarchitecture* (2014), pp. 127–139
48. J.S. Miguel, J. Albericio, A. Moshovos, N.E. Jerger, Doppelgänger: a cache for approximate computing, in *2015 48th Annual IEEE/ACM International Symposium on Microarchitecture (MICRO)* (2015), pp. 50–61
49. A. Yazdanbakhsh, G. Pekhimenko, B. Thwaites, H. Esmaeilzadeh, O. Mutlu, T.C. Mowry, RFVP: rollback-free value prediction with safe-to-approximate loads. ACM Trans. Archit. Code Optim. **12**(4), 1–26 (2016)
50. Universal approximation theorem. https://en.wikipedia.org/wiki/Universalapproximationtheorem. Accessed 07 Sept 2021
51. A. Yazdanbakhsh, J. Park, H. Sharma, P. Lotfi-Kamran, H. Esmaeilzadeh, Neural acceleration for GPU throughput processors, in *Proceedings of the 48th International Symposium on Microarchitecture*. MICRO-48 (Association for Computing Machinery, Waikiki, 2015), pp. 482–493
52. H. Esmaeilzadeh, A. Sampson, L. Ceze, D. Burger, Neural acceleration for general-purpose approximate programs, in *2012 45th Annual IEEE/ACM International Symposium on Microarchitecture* (IEEE, Piscataway, 2012), pp. 449–460

53. T. Moreau, M. Wyse, J. Nelson, A. Sampson, H. Esmaeilzadeh, L. Ceze, M. Oskin, SNNAP: approximate computing on programmable SoCs via neural acceleration, in *2015 IEEE 21st International Symposium on High Performance Computer Architecture (HPCA)* (IEEE. 2015), pp. 603–614

54. R. Nath, D. Tullsen, Chapter 18—accurately modeling GPGPU frequency scaling with the CRISP performance model, in *Advances in GPU Research and Practice*, ed. by H. Sarbazi-Azad. Emerging Trends in Computer Science and Applied Computing (Morgan Kaufmann, Boston, 2017), pp. 471–505

55. S. Herbert, D. Marculescu, Analysis of dynamic voltage/frequency scaling in chip-multiprocessors, in *Proceedings of the 2007 International Symposium on Low Power Electronics and Design (ISLPED'07)* (IEEE. 2007), pp. 38–43

56. H. David, C. Fallin, E. Gorbatov, U.R. Hanebutte, O. Mutlu, Memory power management via dynamic voltage/frequency scaling, in *Proceedings of the 8th ACM International Conference on Autonomic Computing* (2011), pp. 31–40

57. M. Shoushtari, A. BanaiyanMofrad, N. Dutt, Exploiting partially- forgetful memories for approximate computing. IEEE Embed. Syst. Lett. **7**(1), 19–22 (2015)

58. A. Rahimi, A. Ghofrani, K.-T. Cheng, L. Benini, R.K. Gupta, Approximate associative memristive memory for energy-efficient GPUs, in *2015 Design, Automation Test in Europe Conference Exhibition (DATE)* (2015), pp. 1497–1502

59. P.K. Krause, I. Polian, Adaptive voltage over-scaling for resilient applications, in *2011 Design, Automation Test in Europe* (2011), pp. 1–6

60. J. Schlachter, V. Camus, K.V. Palem, C. Enz, Design and applications of approximate circuits by gate-level pruning. IEEE Trans. Very Large Scale Integr. Syst. **25**(5), 1694–1702 (2017)

61. V. Sze, Y.-H. Chen, T.-J. Yang, J.S. Emer, Efficient processing of deep neural networks: a tutorial and survey. Proc. IEEE **105**(12), 2295–2329 (2017)

62. Z. Yang, A. Jain, J. Liang, J. Han, F. Lombardi, Approximate XOR/XNOR-based adders for inexact computing, in *2013 13th IEEE International Conference on Nanotechnology (IEEE-NANO 2013)* (2013), pp. 690–693

63. M. Shafique, W. Ahmad, R. Hafiz, J. Henkel, A low latency generic accuracy configurable adder, in *2015 52nd ACM/EDAC/IEEE Design Automation Conference (DAC)* (2015), pp. 1–6

64. A. Becher, J. Echavarria, D. Ziener, S. Wildermann, J. Teich, A LUT-based approximate adder, in *2016 IEEE 24th Annual International Symposium on Field-Programmable Custom Computing Machines (FCCM)* (2016), pp. 27–27

65. B.S. Prabakaran, S. Rehman, M.A. Hanif, S. Ullah, G. Mazaheri, A. Kumar, M. Shafique, De- MAS: an efficient design methodology for building approximate adders for FPGA-based systems, in *2018 Design, Automation Test in Europe Conference Exhibition (DATE)* (2018), pp. 917–920

66. H. Jiang, F.J.H. Santiago, H. Mo, L. Liu, J. Han, Approximate arithmetic circuits: a survey, characterization, and recent applications. Proc. IEEE **108**(12), 2108–2135 (2020)

67. C.S. Wallace, A suggestion for a fast multiplier. IEEE Trans. Electron. Comput. **1**, 14–17 (1964)

68. L. Dadda, Some schemes for parallel multipliers. Alta Frequenza **34**, 349–356 (1965)

69. G. Zervakis, K. Tsoumanis, S. Xydis, D. Soudris, K. Pekmestzi, Design-efficient approximate multiplication circuits through partial product perforation. IEEE Trans. Very Large Scale Integr. Syst. **24**(10), 3105–3117 (2016)

70. H.R. Mahdiani, A. Ahmadi, S.M. Fakhraie, C. Lucas, Bio-inspired imprecise computational blocks for efficient VLSI implementation of soft- computing applications. IEEE Trans. Circuits Syst. I: Regul. Pap. **57**(4), 850–862 (2010)

71. K.Y. Kyaw, W.L. Goh, K.S. Yeo, Low-power highspeed multiplier for error-tolerant application, in *2010 IEEE International Conference of Electron Devices and Solid-State Circuits (EDSSC)* (2010), pp. 1–4

72. S. Hashemi, R.I. Bahar, S. Reda, DRUM: a dynamic range unbiased multiplier for approximate applications, in *2015 IEEE/ACM International Conference on Computer-Aided Design (ICCAD)* (2015), pp. 418–425

73. H. Saadat, H. Bokhari, S. Parameswaran, Minimally biased multipliers for approximate integer and floating-point multiplication. IEEE Trans. Comput.-Aided Design Integr. Circuits Syst. **37**(11), 2623–2635 (2018)

74. H. Saadat, H. Javaid, A. Ignjatovic, S. Parameswaran, Realm: reduced-error approximate log-based integer multiplier, in *2020 Design, Automation & Test in Europe Conference & Exhibition (DATE)* (IEEE, Piscataway, 2020), pp. 1366–1371

75. K. Bhardwaj, P.S. Mane, J. Henkel, Power- and area efficient approximate Wallace Tree Multiplier for error-resilient systems, in *Fifteenth International Symposium on Quality Electronic Design* (2014), pp. 263–269

76. S. Rehman, W. El-Harouni, M. Shafique, A. Kumar, J. Henkel, J. Henkel, Architectural-space exploration of approximate multipliers, in *2016 IEEE/ACMInternational Conference on Computer-Aided Design (ICCAD)* (IEEE, Piscataway, 2016), pp. 1–8

77. M.S. Ansari, H. Jiang, B.F. Cockburn, J. Han, Low-power approximate multipliers using encoded partial products and approximate compressors. IEEE J. Emerg. Sel. Topics Circuits Syst. **8**(3), 404–416 (2018)

78. C.-H. Lin, I.-C. Lin, High accuracy approximate multiplier with error correction, in *2013 IEEE 31st International Conference on Computer Design (ICCD)* (2013), pp. 33–38

79. S. Venkatachalam, S.-B. Ko, Design of power and area efficient approximate multipliers. IEEE Trans. Very Large Scale Integr. Syst. **25**(5), 1782–1786 (2017)

80. Y. Fu, W. Ephrem, V. Kathail, Embedded vision with int8 optimization on Xilinx devices, in *WP490 (v1. 0.1)*, vol. 19 (2017), p. 15

81. A. Putnam et al., A reconfigurable fabric for accelerating large-scale datacenter services, in *Proceeding of the 41st Annual International Sympo sium on Computer Architecuture (ISCA)*. Selected as an IEEE Micro TopPick (IEEE Press, 2014), pp. 13–24

82. Amazon EC2 F1 Instances. https://aws.amazon.com/ec2/instancetypes/f1/. Accessed 20 Aug 2021

83. Acceleration in the data center—Intel® FPGA. https://www.intel.de/content/www/de/de/data-center/products/programmable/overview.html. Accessed 20 Aug 2021

84. Overview—Alibaba Cloud. https://www.alibabacloud.com/help/docdetail/108504.htm. Accessed 20 Aug 2021

85. Xilinx powers Huawei FPGA Accelerated cloud server (2017). https://www.xilinx.com/news/press/2017/xilinx-powers-huawei-fpga-accelerated-cloud-server.html. Accessed 20 Aug 2021

86. M. Shafique, W. Ahmad, R. Hafiz, J. Henkel, A low latency generic accuracy configurable adder, in *Proceedings of the 52nd Annual Design Automation Conference*. DAC '15 (Association for Computing Machinery, San Francisco, 2015)

87. V. Mrazek, M.A. Hanif, Z. Vasicek, L. Sekanina, M. Shafique, AutoAx: An automatic design space exploration and circuit building methodology utilizing libraries of approximate components, in: *Proceedings of the 56th Annual Design Automation Conference 2019*. DAC '19 (Association for Computing Machinery, Las Vegas, 2019)

88. S. Ullah, S. Rehman, M. Shafique, A. Kumar, High-performance accurate and approximate multipliers for FPGA-based hardware accelerators. IEEE Trans. Comput.-Aided Design Integr. Circuits Syst. **41**(2), 211–224

89. S. Ullah, H. Schmidl, S.S. Sahoo, S. Rehman, A. Kumar, Area-optimized accurate and approximate softcore signed multiplier architectures. IEEE Trans. Comput. **70**(3), 384–392 (2021)

90. S. Ullah, T.D.A. Nguyen, A. Kumar, Energy-efficient low-latency signed multiplier for FPGA-based hardware accelerators. IEEE Embed. Syst. Lett. **13**(2), 41–44 (2021)

91. A.R. Baranwal, S. Ullah, S.S. Sahoo, A. Kumar, ReLAccS: a multilevel approach to accelerator design for reinforcement learning on FPGA-based systems. IEEE Trans. Comput.- Aided Design Integr. Circuits Syst. **40**(9), 1754–1767 (2021)

92. S. Ullah, S. Rehman, B.S. Prabakaran, F. Kriebel, M.A. Hanif, M. Shafique, A. Kumar, Area-optimized low-latency approximate multipliers for FPGA-based hardware accelerators, in *Proceedings of the 55th Annual Design Automation Conference*. DAC '18 (Association for Computing Machinery, San Francisco, 2018),

93. S. Ullah, S.S. Murthy, A. Kumar, SMApproxlib: library of FPGA-based approximate multipliers, in *2018 DAC* (IEEE, Piscataway, 2018), pp. 1–6
94. S. Ullah, S.S. Sahoo, N. Ahmed, D. Chaudhury, A. Kumar, AppAxO: designing application-specific approximate operators for FPGA-based embedded systems. ACM Trans. Embed. Comput. Syst. **21**(3), 1–31 (2022)
95. S. Ullah, S.S. Sahoo, A. Kumar, CLAppED: a design framework for implementing cross-layer approximation in FPGA-based embedded systems, in *2021 58th ACM/IEEE Design Automation Conference (DAC)* (2021), pp. 1–6

Chapter 2
Preliminaries

2.1 Introduction

Chapter 1 has introduced the paradigm of approximate computing as a viable solution for designing resource-efficient, low-latency, and energy-efficient hardware accelerators for error-resilient applications. Toward this end, this book contributes by providing FPGA-based various accurate and approximate arithmetic operators and number representation schemes and utilizes various standard optimization techniques to provide optimal design points for various applications. This chapter provides the preliminary information required to understand the various contributions of the book better. Specifically, it provides information about the logic architecture of Xilinx FPGAs in Sect. 2.2. The various multiplication algorithms used to design accurate and approximate multipliers are discussed in Sect. 2.3. Section 2.4 introduces and discusses the various standard statistical error metrics used to characterize the designed approximate operators. To explore the large design space of approximate circuits, we have used multiple standard optimization techniques such as MBO and GA. These algorithms are introduced and discussed in Sect. 2.5. Finally, Sect. 2.6 briefly introduces ANNs and DNNs, which have been used in many experiments in the book.

2.2 Xilinx FPGA Slice Structure

State-of-the-art FPGAs, such as those provided by Xilinx and Intel, utilize 6-input LUTs to implement combinational and sequential circuits [1, 2]. In this book, we have used Xilinx FPGAs for the implementation of all designs.

The Configurable Logic Block (CLB) is the main computational block of FPGAs for implementing any kind of circuits on FPGAs. The CLB of a modern Xilinx FPGA, such as Xilinx UltraScale, consists of one slice having eight 6-input lookup

S. Ullah, A. Kumar, *Approximate Arithmetic Circuit Architectures for FPGA-based Systems*, https://doi.org/10.1007/978-3-031-21294-9_2

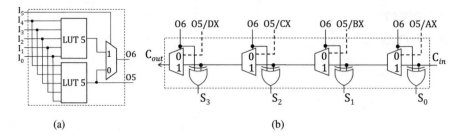

Fig. 2.1 Xilinx FPGA slice structure [1]: (**a**) 6-input LUT, (**b**) 4-bit wide carry chain

tables (referred to as LUT6_2), 8-bit long carry chain, and 16 flip-flops [3]. The same resources are arranged into two slices in a Xilinx 7 series FPGA [1]—which has been used for most of the implementations in this book. As shown in Fig. 2.1a, a LUT6_2 can be used to implement either a single 6-input combinational circuit or two 5-input combinational circuits. For the configuration of LUT6_2, a 64-bit INIT value is assigned to it. This INIT value denotes all the input combinations of LUT6_2 for which a "1" is received at the output. For example, an INIT value of *0000000000000002(hex)* for *LUT6_2* defines to produce outputs *O5 = 1 & O6 = 0* for input combination *100001*. Besides the implementation of combinational functions, these 6-input LUTs are also used for controlling the associated carry chain, as shown in Fig. 2.1b. The carry chain implements a carry-lookahead adder using O5 as the carry-generate signal and O6 as the carry-propagate signal as described by Eqs. 2.1 and 2.2. However, O5 can be bypassed by the external IX signal for providing the carry-generate signal. The input carry, "CIN", can be either assigned to constant "0/1" or to "COUT" of another carry chain from a different slice.

$$S_i = P_i \oplus C_i \tag{2.1}$$

$$C_{i+1} = G_i \cdot \overline{P_i} + P_i \cdot C_i \tag{2.2}$$

2.3 Multiplication Algorithms

In this book, we have used Baugh-Wooley's and Booth's multiplication algorithms for implementing accurate and approximate signed multipliers. In the following subsections, we provide a brief overview of the two multiplication algorithms.

2.3.1 Baugh-Wooley's Multiplication Algorithm

Compared to unsigned multiplication, all the partial products in a signed multiplication must be properly sign-extended to compute the accurate product. Baugh-Wooley's multiplication algorithm [4] eliminates the need for computing and communicating sign-extension bits by encoding the sign information in the generated partial products. For an $N \times M$ signed multiplier, Eq. 2.3 describes the respective operands in 2's complement representation. Equation 2.4 illustrates the generation of the signed partial products to compute the final product "P". Baugh-Wooley's multiplication algorithm rewrites the negative partial product terms, as described in Eq. 2.5, to eliminate the need for explicit sign-extension bits. The $\overline{a_x b_y}$ term in the equation, $for\ x \in [0 \ldots N - 1]\ and\ y \in [0 \ldots M - 1]$, denotes the 1's complement of the corresponding partial product term. For example, for an 8×8 signed multiplier, Eq. 2.6 represents the signed partial products according to Baugh-Wooley's algorithm.

$$A = -a_{N-1}2^{N-1} + \sum_{n=0}^{N-2} a_n 2^n$$

$$B = -b_{M-1}2^{M-1} + \sum_{m=0}^{M-2} b_m 2^m \tag{2.3}$$

$$P = a_{N-1}b_{M-1}2^{N+M-2} + \sum_{n=0}^{N-2}\sum_{m=0}^{M-2} a_n b_m 2^{n+m} - 2^{N-1}\sum_{m=0}^{M-2} a_{N-1}b_m 2^m$$

$$- 2^{M-1}\sum_{n=0}^{N-2} b_{M-1}a_n 2^n \tag{2.4}$$

$$P = a_{N-1}b_{M-1}2^{N+M-2} + \sum_{n=0}^{N-2}\sum_{m=0}^{M-2} a_n b_m 2^{n+m} + 2^{N-1}\sum_{m=0}^{M-2} \overline{a_{N-1}b_m} 2^m$$

$$+ 2^{M-1}\sum_{n=0}^{N-2} \overline{b_{M-1}a_n} 2^n + 2^{N-1} + 2^{M-1} + 2^{N+M-1} \tag{2.5}$$

$$P_8 = a_7 b_7 2^{14} + \sum_{n=0}^{6}\sum_{m=0}^{6} a_n b_m 2^{n+m} + 2^7 \sum_{m=0}^{6} \overline{a_7 b_m} 2^m + 2^7 \sum_{n=0}^{6} \overline{b_7 a_n} 2^n$$

$$+ 2^8 + 2^{15} \tag{2.6}$$

Table 2.1 Booth's encoding

S. No.	b_{n+1}	b_n	b_{n-1}	BE	s	c	z
0	0	0	0	0	0	0	1
1	0	0	1	1	0	0	0
2	0	1	0	1	0	0	0
3	0	1	1	2	1	0	0
4	1	0	0	$\bar{2}$	1	1	0
5	1	0	1	$\bar{1}$	0	1	0
6	1	1	0	$\bar{1}$	0	1	0
7	1	1	1	0	0	0	1

2.3.2 Booth's Multiplication Algorithm

Booth's multiplication algorithm reduces the number of partial products to enhance the performance of a multiplier. A radix-4 Booth's multiplier halves the total number of partial products for an M × N signed multiplier. Equation 2.7 shows the 2's complement representations of *multiplicand A* and *multiplier B*, and the corresponding radix-4 booth's multiplication is summarized in Eq. 2.8.

$$A = -a_{M-1}2^{M-1} + \cdots + a_2 2^2 + a_1 2^1 + a_0$$
$$B = -b_{N-1}2^{N-1} + \cdots + b_2 2^2 + b_1 2^1 + b_0 \tag{2.7}$$

$$A \cdot B = \sum_{n=0}^{N/2} A \cdot BE_{2n} 2^{2n} \tag{2.8}$$

$$where \quad BE_{2n} = -2b_{2n+1} + b_{2n} + b_{2n-1}$$

The values of Booth's encoding (BE) in Eq. 2.8 are in the range of $\pm 0, \pm 1, \pm 2$ and can be computed as shown in Table 2.1. A partial product is shifted left if BE = 2 (denoted by $s = 1$). Similarly, for a negative value of BE (denoted by $c = 1$), the 2's complement of the corresponding partial product is calculated by initially taking 1's complement of the partial product and adding a '1' to the least significant bit (LSB) position. For BE = 0 (denoted by $z = 1$), the corresponding partial product is replaced by a string of zeros.

2.3.3 Sign Extension for Booth's Multiplier

As discussed previously, all the partial products must be adequately sign-extended in a signed multiplication before reducing them to a final product. As shown by rows number 4, 5, and 6 in Table 2.1, Booth's encoding can result in negative partial products. However, the correct sign of a partial product in Booth's multiplier

Table 2.2 Sign extension for Booth's multiplier

b_{n+1}	b_n	b_{n-1}	BE	MSB multiplicand	SE
0	0	0	0	0	0
0	0	0	0	1	0
0	0	1	1	0	0
0	0	1	1	1	1
0	1	0	1	0	0
0	1	0	1	1	1
0	1	1	2	0	0
0	1	1	2	1	1
1	0	0	$\overline{2}$	0	1
1	0	0	$\overline{2}$	1	0
1	0	1	$\overline{1}$	0	1
1	0	1	$\overline{1}$	1	0
1	1	0	$\overline{1}$	0	1
1	1	0	$\overline{1}$	1	0
1	1	1	0	0	0
1	1	1	0	1	0

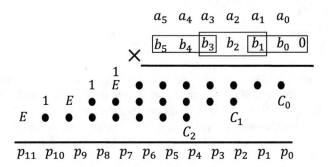

Fig. 2.2 Sign extension of partial products in a 6×6 Booth's multiplier [5]

depends on the BE and the most significant bit (MSB) of the multiplicand. If the MSB of the multiplicand is "0" and the BE is either a positive number or "0", the partial product will be 0-extended. Similarly, if the MSB of the multiplicand is "1" and the BE is a nonzero positive number, the partial product will be extended with "1". A complete list of all possible sign extension cases, denoted by "SE", is presented in Table 2.2.

Using Bewick's method in [5], sign extension information can be efficiently encoded in the generated partial products. Figure 2.2 shows an example of Bewick's sign extension technique for a 6×6 Booth's multiplier. The C_0, C_1, and C_2 will be "1" for negative partial products to represent them in 2's complement notation. The "E" bits are the complement of corresponding "SE" bits in Table 2.2. This technique significantly reduces the sign extension of each partial product row to a maximum of two more bit positions.

2.4   Statistical Error Metrics

Statistical error metrics are utilized to characterize the output quality of an approximate circuit. These metrics denote the magnitude and frequency of errors manifested in the approximate output (O_{App}) by comparing it with accurate output (O_{Acc}). This book employs some of the commonly utilized error metrics to characterize the proposed approximate circuits [6–9]. Here we provide an overview of these metrics.

- *Maximum error magnitude:* As defined in Eq. 2.9, it is the maximum absolute difference between O_{Acc} and O_{App} for all possible input combinations:

$$Maximum\ error = \max_{\forall inputs} (|O_{Acc} - O_{App}|) \tag{2.9}$$

- *Average absolute error:* It is the average of the absolute difference between O_{Acc} and O_{App} over all input combinations. Equation 2.10 defines the metric mathematically:

$$Average\ absolute\ error = \frac{1}{N} \sum_{i=0}^{N} |O_{Acc_i} - O_{App_i}| \tag{2.10}$$

- *Average error:* In many cases for error-resilient applications, the errors produced for various input combinations cancel each other out. To this end, the average error metric, as defined in Eq. 2.11, estimates the final error. It is the average of the difference between O_{Acc} and O_{App} over all input combinations:

$$Average\ error = \frac{1}{N} \sum_{i=0}^{N} (O_{Acc} - O_{App}) \tag{2.11}$$

- *Average absolute relative error:* As defined in Eq. 2.12, this metric characterizes the error magnitude with respect to the accurate answer. For example, for $O_{Acc} = 1$, the error magnitude of 2 is very high. However, the same error magnitude of 2 for $O_{Acc} = 1000$ is negligible. The error metric is calculated by computing the ratios of absolute errors and corresponding accurate answers and taking the average over all input combinations:

$$Average\ absolute\ relative\ error = \frac{1}{N} \sum_{i=0}^{N} |\frac{O_{Acc} - O_{App}}{O_{Acc}}| \tag{2.12}$$

- *Average relative error:* As described in Eq. 2.13, it is computed by taking the average of all ratios of differences between O_{Acc} and O_{App} and the corresponding O_{Acc}:

$$Average \; relative \; error = \frac{1}{N} \sum_{i=0}^{N} (\frac{O_{Acc} - O_{App}}{O_{Acc}}) \qquad (2.13)$$

- *Maximum absolute relative error:* As defined in Eq. 2.14, this metric computes the maximum absolute relative error for all input combinations. Together with the average absolute relative error, this metric helps in estimating the overall error distribution of an operator. For example, for an operator, a high value of maximum absolute relative error and a comparatively lower value of average absolute relative error show that most of the outputs of the operator have high output accuracy by generating lower absolute relative error values. Further, one should expect a few output values with high absolute relative error values:

$$Maximum \; absolute \; relative \; error = \max_{\forall \; inputs} (|\frac{O_{Acc} - O_{App}}{O_{Acc}}|) \qquad (2.14)$$

- *Mean squared error:* As described in Eq. 2.15, this metric is computed by taking the average of all squares of differences between O_{Acc} and O_{App} for all input combinations:

$$Mean \; squared \; error = \frac{1}{N} \sum_{i=0}^{N} (O_{Acc} - O_{App})^2 \qquad (2.15)$$

- *Error probability:* As defined in Eq. 2.16, this error metric computes the total error occurrences for all input combinations:

$$Error \; probability = \frac{Total \; number \; of \; outputs \; where \; O_{Acc} \neq O_{App}}{Total \; number \; of \; outputs} \qquad (2.16)$$

- *Maximum error probability:* As described in Eq. 2.17, this error metric defines the total occurrences of the maximum error magnitude for all input combinations:

$$Maximum \; error \; probability = \frac{Maximum \; error \; occurrences}{Total \; number \; of \; outputs} \qquad (2.17)$$

2.5   Design Space Exploration and Optimization Techniques

Design optimization is a common and increasingly explored research challenge in various domains of scientific and engineering applications. It focuses on finding (or tuning) the values of various design parameters that provide the maximal or minimal output values of an objective function. However, in many cases, more than one objective function needs to be optimized for a single design. For example, to design a hardware accelerator for an application, a designer is usually confronted with various design constraints such as minimal output accuracy degradation, desired throughput, and maximum energy consumption of the implementation. To satisfy these conflicting objective functions, the designer has to find the optimal configuration of various design choices, such as the types of instructions to be executed, the precision of operands, and the accuracy of intermediate computations. For example, reducing the precision of operands can reduce the overall energy consumption of the implementation at the cost of degraded output quality. Similarly, the utilization of approximate operators can help improve the accelerator's performance by trading the corresponding output quality. Such design optimization problems are commonly known as multi-objective optimization problems. Mathematically, for an n-dimensional design space $\mathcal{X} \subseteq \mathbb{R}^n$ and m-dimensional vector function $\overrightarrow{f}$ where $\overrightarrow{f} : \mathcal{X} \mapsto \mathcal{Z}$ and $\mathcal{Z} \subseteq \mathbb{R}^m$, the multi-objective optimization problem can be formulated as [10]

$$\overrightarrow{x}^* = \text{argmax}_{\overrightarrow{x} \in \mathcal{X}}(f_1(\overrightarrow{x}), f_2(\overrightarrow{x}), \ldots f_m(\overrightarrow{x})) \tag{2.18}$$

where $\overrightarrow{x}^*$ represents the design configuration that provides the maximal (or minimal) value of a multi-objective design space.

However, in many cases, the individual objectives in a multi-objective optimization problem have contradictory characteristics, and the optimal values for each objective function are in different regions of the whole design space. For instance, for the hardware accelerator example, as abovementioned, the objective function for energy consumption will try to utilize the minimum available precision (bit width) for operands. In contrast, the objective function for output accuracy will opt for maximum available precision. For multi-objective optimization problems, the optimal values for one objective function may be in the non-optimal region for another objective function. In such scenarios, multi-objective optimization aims to provide a set of Pareto (non-dominated) design configurations that provide feasible trade-offs between the different objective functions. To this end, various optimization algorithms are available that can assist in the efficient exploration of large design spaces and the identification of feasible design points. In this book, we have utilized Genetic Algorithm (GA) and Multi-objective Bayesian Optimization (MBO) as optimization algorithms for the design space exploration of approximate circuits.

2.5.1 Genetic Algorithm

GA-based optimization is a stochastic search technique inspired by natural selection-based genetic evolution [11, 12]. In this technique, a pool of solutions (current population) is utilized for identifying the feasible solutions (the fittest) to produce the next generation of solutions by employing genetic processes. This

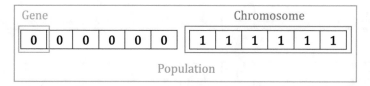

Fig. 2.3 Gene, chromosome, and population in Genetic Algorithm

process of producing the next generations of solutions can be repeated for a fixed number of iterations or until further iterations produce the next generations without significant improvements. Each candidate solution in the population is represented by an n-bit string (chromosome), as described in Fig. 2.3. To produce the next generations of solutions, GA iterates over the following four processes:

- *Fitness function evaluation:* All the candidate solutions in a population are evaluated using an objective function to determine the fitness (cost) of the solution. For example, for designing an energy-efficient approximate accelerator, the objective functions can be determining critical path delay and power consumption of various implementations.
- *Selection of candidate solutions:* In this phase, a number of solutions are selected based on the fitness (score) of the candidate solutions to produce the next generation of population. For this purpose, various techniques, such as roulette wheel- and rank-based selection, can be employed for selecting a candidate. In this book, we have used the tournament-based selection method for our GA-based exploration. In this method, $\mathcal{K}$ solutions ($\mathcal{K}$ being a hyperparameter) are picked randomly, and the fittest member among them is selected for further processing. During this process, a single solution can be used multiple times to produce new solutions.
- *Crossover:* In this step, the selected solutions are grouped in pairs (parents) to produce two new solutions (children). For this purpose, a crossover point C is selected randomly where $0 < C < length\ of\ the\ chromosome$. The two parents' chromosome values (genes) are swapped together across the crossover point to produce two new solutions (children). For example, Fig. 2.4 shows an example of two parent solutions along with the two new solutions for $C = 3$. The example shows a one-point crossover; however, other crossover schemes, such as two-point crossover, are also possible. Further, crossover is a hyperparameter-controlled probabilistic process; therefore, some parents can be transferred to the new solutions without any alteration.
- *Mutation:* To avoid premature convergence of the population and maintain diversity in the solutions, GA performs the mutation process. In this process, some gene values in the new solutions are flipped. It is also a hyperparameter-controlled probabilistic process.

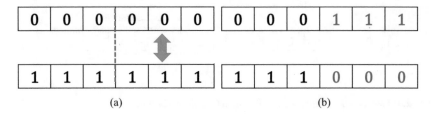

Fig. 2.4 Genetic Algorithm crossover. (**a**) Parent solutions. (**b**) New solutions

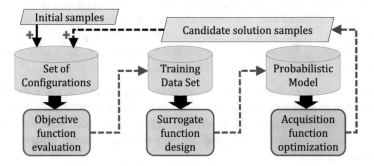

Fig. 2.5 Bayesian optimization flow

2.5.2 Bayesian Optimization

GA-based optimization usually involves evaluating a large number of configurations using the objective function to find a feasible set of configurations. In some cases, the objective function has high computational complexity and requires significant time and resources to evaluate even a single configuration. For example, the synthesis and implementation of an accelerator for an error-resilient application, such as an artificial neural network, can take several minutes to evaluate the performance impact of utilizing a set of approximate multipliers. For optimizing such complex objective functions, Bayesian optimization provides a more directed search method and utilizes *surrogate functions* to reduce the evaluation time of a configuration [10, 13]. Figure 2.5 describes the overall flow of Bayesian optimization. It consists of the following three stages:

- *Objective function evaluation:* It involves the actual evaluation of an initial set of random configurations using the true objective function. The output of this stage forms the training set for the next stage.
- *Surrogate function design:* It is a probabilistic model utilized to approximate the objective function by mapping the inputs to the estimated outputs. Different techniques, such as the tree Parzen estimator and the random forest regression, can be utilized for designing the surrogate function. However, among these available options, the Gaussian process is a commonly utilized technique [14], and we have also used it for the surrogate function design in this book. It can

approximate a diverse set of functions and provides smooth transitions between outputs as new training data is made available. In an MBO, separate surrogate functions are utilized for every individual objective function.

• *Acquisition function optimization:* In this step, the acquisition function samples the design space using some predefined technique, such as random search, and utilizes the surrogate function for finding the corresponding approximate outputs (fitness value) of the new configurations. The evaluated points are ranked using techniques such as the probability of improvement, expected improvement and lower confidence bound. These techniques mainly differ in their tolerance for the exploitative and explorative search of the design space. An exploitative search technique tries to cover samples (new configurations) where the surrogate function estimates a good value of the objective function. In contrast, the explorative technique also provides the samples where the surrogate function provides a low objective function value. The selected samples are then provided to the true objective function to compute their actual values. Finally, the newly computed values are added to the original training dataset for the next iteration of the algorithm. The acquisition function guides the search direction of the whole algorithm in finding the configurations that provide the maximal or minimal values of the objective function.

2.6 Artificial Neural Networks

Machine learning is a branch of Artificial Intelligence (AI) that enables algorithms to analyze, extract patterns, and learn from data without explicitly being programmed. The well-known Artificial Neural Networks (ANNs) are a class of machine learning algorithms inspired by the working of a human brain where one neuron signals a set of other neurons. An artificial neuron (commonly known as a perceptron) is the building block of defining different types of layers and different types of artificial neural networks. As shown in Fig. 2.6, a perceptron receives multiple inputs, performs a weighted sum of these inputs, adds a bias to the sum, and finally applies an activation function to produce the output. The activation function introduces nonlinearity into ANNs to enable them to analyze, understand, and infer real-world nonlinear data. For this purpose, the commonly utilized activation functions are the sigmoid function, rectified linear unit (ReLU), and the hyperbolic tangent, as defined in Eq. 2.19:

$$Sigmoid\ function = \frac{1}{1 + e^{-y}}$$

$$ReLU = max(0, y) \tag{2.19}$$

$$Hyperbolic\ tangent = \frac{e^y - e^{-y}}{e^y + e^{-y}}$$

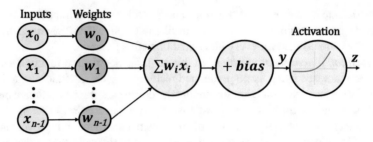

Fig. 2.6 Perceptron: artificial neuron

An ANN containing more than three layers is usually referred to as a Deep Neural Network (DNN). A DNN can contain layers of different types, such as fully connected (dense), convolution, and pooling layers. These layers differ in their computations on the input data. For example, every perceptron in a dense layer receives inputs from all neurons in the previous layer, whereas the neurons in a convolution layer preserve spatial correlation between input values by utilizing filters to focus on only a part of the input data (receptive field). Similarly, the pooling layers in a DNN are used to subsample and reduce the size of feature maps. The frequently employed pooling function for this purpose is Max-pooling which computes the maximum feature "z" according to the window size of the neurons in the layer. For an $\frac{n}{2} \times \frac{n}{2}$ window size, the output of a neuron in Max-pooling layer is computed as in Eq. 2.20:

$$z_{out} = max(z_1, z_2, z_3, \ldots, z_n) \tag{2.20}$$

The utilization of ANNs for any application consists of multiple design decisions and stages. The first step is to prepare different input data sets for training, validating, and testing a network. For example, ImageNet is a benchmark dataset for testing the efficacy of various DNNs for the image classification task [15]. Similarly, for speech processing, LibriSpeech is a commonly utilized dataset [16]. The second step is the actual design of the ANN. It involves different design decisions such as the number and types of layers, types of activation functions, and deciding the loss function, learning rate, optimizer, and dropout rate for training the network. For example, Fig. 2.7 shows an example of a state-of-the-art Convolutional Neural Network (CNN) utilized to classify the ImageNet dataset. In this book, we have used MLPs—ANNs deploying only dense layers—as test applications for evaluating the efficacy of different contributions.

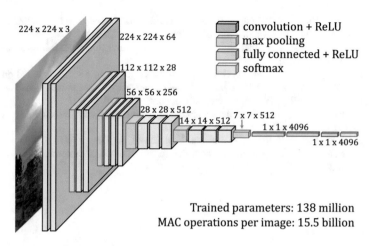

Trained parameters: 138 million
MAC operations per image: 15.5 billion

Fig. 2.7 VGG-16 DNN for classification of ImageNet dataset [17, 18]

References

1. Xilinx. 7 Series FPGAs Configurable Logic Block. (2016) https://www.xilinx.com/support/documentation/user_guides/ug4747SeriesCLB.pdf
2. Intel® Stratix® 10 Logic Array Blocks and Adaptive Logic Modules User Guide (2020). https://www.intel.com/content/dam/www/programmable/us/en/pdfs/literature/hb/stratix-10/ug-s10-lab.pdf
3. Xilinx. UltraScale Architecture Configurable Logic Block (2017). https://www.xilinx.com/support/documentation/user_guides/ug574-ultrascaleclb.pdf
4. C.R. Baugh, B.A. Wooley, A two's complement parallel array multiplication algorithm. IEEE Trans. Comput. **C-22**(12), 1045–1047 (1973)
5. G.W. Bewick, Fast multiplication: Algorithms and implementation. PhD Thesis. Stanford University, 1994
6. P. Stanley-Marbell et al., Exploiting errors for efficiency: a survey from circuits to applications. ACM Comput. Surv. **53**(3), 1–39 (2020)
7. S. Rehman, W. El-Harouni, M. Shafique, A. Kumar, J. Henkel, J. Henkel, Architectural-space exploration of approximate multipliers, in *2016 IEEE/ACMInternational Conference on Computer-Aided Design (ICCAD)* (IEEE, Piscataway, 2016), pp. 1–8
8. H. Jiang, F.J.H. Santiago, H. Mo, L. Liu, J. Han, Approximate arithmetic circuits: a survey, characterization, and recent applications. Proc. IEEE **108**(12), 2108–2135 (2020)
9. S. Hashemi, R. Iris Bahar, S. Reda, DRUM: a dynamic range unbiased multiplier for approximate applications, in *2015 IEEE/ACM International Conference on Computer-Aided Design (ICCAD)* (2015), pp. 418–425
10. P.P. Galuzio, E.H. de Vasconcelos Segundo, L. dos Santos Coelho, V.C. Mariani, MOBOpt—multiobjective Bayesian optimization. SoftwareX **12**, 100520 (2020)
11. J. Carr, An introduction to genetic algorithms. Senior Project **1**(40), 7 (2014)
12. F.-A. Fortin, F.-M. De Rainville, M.-A. Gardner, M. Parizeau, C. Gagné, DEAP: evolutionary algorithms made easy. J. Mach. Learn. Res. **13**, 2171–2175 (2012)
13. J. Močkus, On Bayesian methods for seeking the extremum, in *Optimization Techniques IFIP Technical Conference Novosibirsk, July 1–7, 1974*, ed. by G.I. Marchuk (Springer, Berlin, 1975), pp. 400–404

14. J. Bergstra, R. Bardenet, Y. Bengio, B. Kégl, Algorithms for hyper-parameter optimization, in *Advances in Neural Information Processing Systems*, vol. 24 (2011)
15. O. Russakovsky et al., Imagenet large scale visual recognition challenge. Int. J. Comput. Vis. **115**(3), 211–252 (2015)
16. V. Panayotov, G. Chen, D. Povey, S. Khudanpur, Librispeech: an asr corpus based on public domain audio books, in *2015 IEEE International Conference on Acoustics, Speech and Signal Processing (ICASSP)* (IEEE, Piscataway, 2015), pp. 5206–5210
17. K. Simonyan, A. Zisserman, Very deep convolutional networks for large-scale image recognition (2014). arXiv preprint arXiv:1409.1556
18. S. Nambi, S. Ullah, S.S. Sahoo, A. Lohana, F. Merchant, A. Kumar, ExPAN(N)D: exploring posits for efficient artificial neural network design in FPGA-based systems. IEEE Access **9**, 103691–103708 (2021)

Chapter 3
Accurate Multipliers

3.1 Introduction

Multiplication is one of the basic arithmetic operations, used extensively in the domain of digital signal, image processing, and ML applications. FPGA vendors, such as Xilinx and Intel, provide DSP blocks to achieve fast multipliers [1, 2]. Despite the high performance offered by these DSP blocks, their usage might not be efficient in terms of overall performance and resource requirements for some applications. For example, Table 3.1 compares the CPDs and LUTs utilization of two different implementations of Reed-Solomon and JPEG encoders[1] for Virtex-7 series FPGA using Xilinx Vivado tool. The routing delay caused by the fixed location of the allocated DSP blocks has resulted in higher CPD for DSP-based implementation of Reed-Solomon encoder. For small applications, it may be possible to perform manual floorplanning to optimize an application's overall performance. However, for complex applications with contending requirements for FPGA resources, it may not be possible to optimize the placement of required FPGA resources manually to enhance performance gains. Similarly, the implementation of the JPEG encoder shows a large number of DSP blocks (56% of the total available DSP blocks) utilization. Such applications can exhaust the available DSP blocks, making them less available/unavailable for performance-critical operations of other applications executing concurrently on the same FPGA, and thereby necessitating the LUT-based multipliers. Moreover, the utilization of DSP blocks having $M \times M$ multipliers for obtaining $Y \times Y$ and $Z \times Z$ multipliers, where $M > Y$ and $M < Z$, can degrade the performance of overall implementations, as elaborated by our experimental results in Sect. 3.9. Therefore, the orthogonal approach of having logic-based soft multipliers along with DSP blocks is important for obtaining overall performance gains in different implementation scenarios. That is why Xilinx and

---

[1] Source codes from *https://opencores.org/projects*.

© The Author(s), under exclusive license to Springer Nature Switzerland AG 2023
S. Ullah, A. Kumar, *Approximate Arithmetic Circuit Architectures for FPGA-based Systems*, https://doi.org/10.1007/978-3-031-21294-9_3

Table 3.1 Comparison of logic vs DSP blocks based implementations for Reed-Solomon decoder and JPEG encoder

Design	DSP blocks enabled			DSP blocks disabled		
	CPD [ns]	LUTs	DSPs	CPD [ns]	LUTs	DSPs
Reed-Solomon dec.	4.68	2797	22	4.47	2839	0
JPEG enc.	8.85	14,780	631	9.88	71,362	0

Intel also provide logic-based soft multipliers [3, 4]. However, in this chapter, we show that these soft multiplier IP cores for FPGAs still need better designs to provide high-performance and resource efficiency.

Towards this end, we have explored various multiplication algorithms and utilized various FPGA-specific optimization techniques to propose various accurate unsigned and signed multipliers. Our proposed accurate multipliers utilize efficient techniques for partial product encoding and their reduction. The following are the key contributions of this chapter.

3.2 Contributions

- *An Accurate Unsigned Multiplier Design:* Utilizing the 6-input LUTs and associated fast carry chains of the state-of-the-art FPGAs, we present a scalable, area-optimized, and reduced latency architecture of accurate unsigned multiplier, denoted as *Acc*.
- *Accurate Signed Multiplier:* Utilizing Baugh-Wooley's multiplication algorithm [5], we extend the unsigned multiplier design to implement an area-optimized signed multiplier design, referred to as *Mult-BW*.
- Utilizing radix-4 Booth's multiplication algorithm [6]: We present an area-optimized and energy-efficient signed multiplier design, denoted as *Booth-Opt*. Our implementation efficiently utilizes the LUTs and carry chains of FPGAs to combine the process of partial product generation and their accumulation.
- To reduce the critical path delay of the *Booth-Opt* design, we present a high-performance signed multiplier design, indicated as *Booth-Par*. This design generates all partial products in parallel and then compressor trees to compute the final product.
- We present a constant multiplier design for applications that employ a constant coefficient as one of the operands to multiplication. Compared to a variable multiplier—where both operands are variable—a constant multiplier provides a significant reduction in the overall utilized resources and critical path delay.

The rest of the chapter is organized as follows. Section 3.3 presents an overview of the related state-of-the-art accurate multiplier designs along with their limitations. Section 3.4 presents the design of our proposed accurate unsigned multiplier design followed by the motivation of signed multipliers in Sect. 3.5 and the extension of the unsigned design to a signed multiplier architecture (*Mult-BW*) in Sect. 3.6. We then present three different designs of signed multipliers using Booth's multiplication algorithm in Sect. 3.7. Section 3.8 presents our proposed constant multiplier design for applications where one of the operands is a constant coefficient in multiplication. The implementation results of the various designs are discussed in Sect. 3.9. Finally, Sect. 3.10 concludes the chapter.

3.3 Related Work

Multiplication is often one of the major contributors to energy consumption, critical path delay, and resource utilization of various applications, such as deep neural networks. These effects get more pronounced in FPGA-based designs. However, most of the state-of-the-art designs are done for ASIC-based systems. Further, few FPGA-based designs that exist are primarily limited to unsigned numbers, which require extra circuits to support signed operations. In this section, we provide a summary of relevant state-of-the-art works.

Walters [7, 8] and Kumm et al. [9] have used the modified Booth's algorithm [6] for area-efficient radix-4 multiplier implementations using 6-input LUTs and associated carry chains of Xilinx FPGAs. Their implementations avoid partial products compressor trees and have large critical path delays. Kumm et al. have utilized integer linear programming to explore the modular design for resource-efficient implementation of larger multipliers from smaller sub-multipliers for Xilinx FPGAs [10]. Their technique considers both the DSP blocks and LUTs to implement the sub-multipliers to efficiently utilize the available resources on an FPGA. The authors of[11] have utilized the 6-input LUTs and carry chains of Xilinx FPGAs to implement a set of constant-coefficient multipliers for DNNs. These multipliers are based on the *shift*, *add*, and *subtract* operations. Parandeh-Afshar et al. have also used the Booth's and Baugh-Wooley's multiplication algorithms [5] for area-efficient multiplier implementation using Altera (now Intel) FPGAs [12]. However, to reduce the effective length of carry chains, their implementation limits the length of the Adaptive Logic Modules (ALMs) to five, which results in the underutilization of the FPGA resources. Moreover, this feature of limiting the carry chain to five ALMs cannot be achieved without wasting resources, with current FPGAs from other vendors, such as those provided by Xilinx [13]. Parandeh-Afshar et al. have also proposed a partial products compressor tree using Altera FPGAs [14]. However, their proposed implementation of generalized parallel counters (GPCs) underutilizes LUTs in two consecutive ALMs. The authors of [15] have utilized the ALMs of Intel FPGA to propose resource-efficient implementations for smaller multipliers, such as 3×3 and 5×5. Their implementation focuses on the restructuring of the partial product terms to reduce the total number of levels required to compute the final product.

Among other available multiplier design options, the conventional shift-and-add [16], serial and serial/parallel multipliers address the low area requirements but offer very high critical path delays. The commonly used Wallace [17] and Dadda [18] design-based parallel multipliers have high area requirements for achieving low output latencies by parallel addition of partial products. Considering the characteristics of the Booth and Wallace/Dadda multiplier schemes, a fast hybrid multiplier architecture using radix-4 recoding has been proposed in [19] for ASIC-based systems.

3.4   Unsigned Multiplier Architecture

The proposed implementation of unsigned multiplier, referred to as *Acc*, is based on the basic method of multiplying two multi-bit numbers $A_{(N\text{-bits})}$ and $B_{(M\text{-bits})}$, as shown in Fig. 3.1. The multiplication results in the generation of M, N-bit partial products with required shifting. The proposed implementation of multiplier fuses the generation and mutual addition of two consecutive partial product rows into one stage. For an $N_{\text{Multiplicand}} \times M_{\text{Multiplier}}$ multiplier, it results in the concurrent generation of $(N + 2)$-bits long $\left\lceil \frac{M}{2} \right\rceil$ Processed Partial Products (PPPs). *Our automated tool flow* organizes the PPPs in $\left\lceil \frac{M}{6} \right\rceil$ groups. Each group can contain a maximum of three PPPs. Using 6-input LUTs and the associated carry chains, our methodology then deploys either ternary or binary adders for the mutual addition of PPPs in each group. The total number of stages required to find final accurate product is defined by Eq. 3.1:

$$\text{No. of stages} = \left\lceil \log_3 \left(\frac{M}{2} \right) \right\rceil + 1 \tag{3.1}$$

Figure 3.2 exhibits the elemental steps of our proposed implementation to realize the "*Acc*" unsigned multiplier. It includes the following operations:

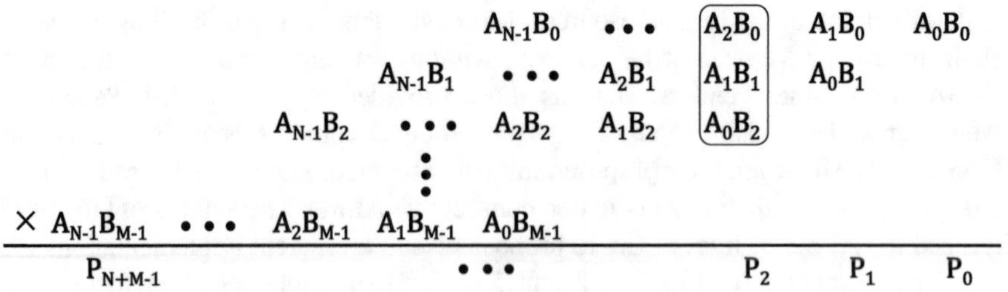

Fig. 3.1 $A_{N\text{-}bit} \times B_{M\text{-}bit}$ basic multiplier design

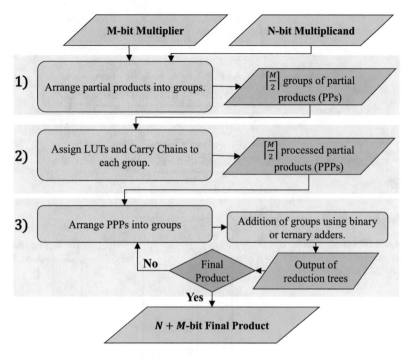

Fig. 3.2 Proposed design flow of *Acc* multiplier

1. *Organization of Partial Products (PPs)*: We have used the 6-input LUTs for computing the required PPs by performing *AND* operation between every bit of multiplier and multiplicand. However, to enhance the utilization of LUTs, our *automated methodology* groups every two consecutive PPs. Each group contains the second PP shifted left by a single bit position relative to the first PP. Further, the PPs in every group are computed and mutually added in one single step. However, in every group, there are two PP terms, for example, $A_0 B_0$ and $A_{N-1} B_1$ in the first group, which are not added with any other PP term in their respective group. Moreover, due to the limited number of input/output pins of LUTs in modern FPGAs, it is not possible to group more than two PP terms. For example, the generation and addition of PPs $A_2 B_0$, $A_1 B_1$ and $A_0 B_2$, as shown by the blue box in Fig. 3.1, cannot be performed in a single step.
2. *LUTs and Carry Chain Assignment*: In this step, our methodology assigns the 6-input LUTs and the associated carry chains to each group of PPs, as shown by the computational blocks Type-A and Type-B in Fig. 3.3. An instance of a block, either Type-A or Type-B, denotes a 6-input LUT with an associated adder and carry chain cell (CC). Figure 3.4a shows the functionality of the LUT of block Type-A. The output signals O5 and O6 are passed to the corresponding carry chain as carry-generate (G_i) and carry-propagate (P_i) signals, respectively. The

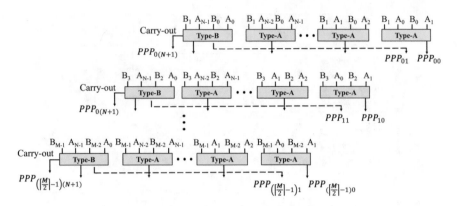

Fig. 3.3 Partial product generation for an $N \times M$ *Acc* multiplier

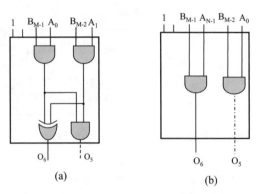

Fig. 3.4 LUTs configuration for *Acc* multiplier. (**a**) LUT of Type-A. (**b**) LUT of Type-B

LUT configuration for block Type-B, in Fig. 3.4b, uses O5 for the computation of the least significant partial product term in each row. The generate signal for the carry chain element corresponding to block Type-B is constant "0" and provided by the external bypass signal (AX-DX), as already described in Fig. 2.1. The associated carry chain uses Eqs. 2.1 and 2.2 for the generation of sum bit (S_i) and carry out (C_{i+1}) bits. The completion of this stage of our implementation results in the generation of ($N + 2$)-bits long $\left\lceil \frac{M}{2} \right\rceil$ Processed Partial Products (PPPs).

3. *Rearrangement and Reduction of PPPs*: Our implementation utilizes ternary and binary adders for reducing PPPs to a final product. Modern FPGAs provide the capability of implementing a ternary adder as a ripple carry adder (a 3:1 compressor for the simultaneous reduction of three partial products) [20]. Our automated methodology arranges the PPPs in multiple groups, with the intention of placing three distinct PPPs in each group. Depending on the value of "M," in Fig. 3.1, a group may have one, two, or three PPPs. Our implementation then utilizes 3:1 (ternary adder) and 2:1 (binary adder) compressors for reducing PPPs in each group. The PPPs reduction phase may produce new partial sums, which are again grouped and passed through 3:1 and 2:1 compressors. This process is

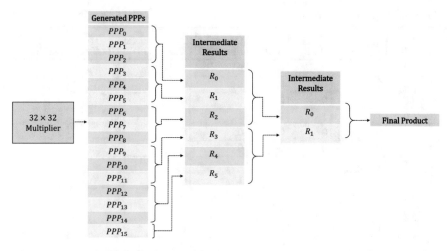

Fig. 3.5 Grouping of PPPs to compute final product for a 32×32 multiplier

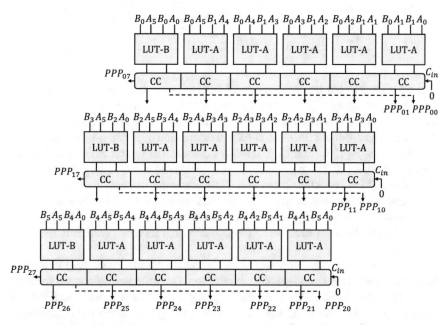

Fig. 3.6 Virtex-7 FPGA slice-based representation of processed partial product generation for a 6×6 *Acc* multiplier

repeated until the final product is obtained. For example, for a 32×32 multiplier, 16 PPPs are generated. The grouping and reduction of these PPPs to compute the final product is described in Fig. 3.5.

Figure 3.6 shows the mapping of the proposed implementation for a 6×6 *Acc* multiplier on Xilinx 7 series FPGAs. However, the same implementation can

Table 3.2 Type-A LUT configuration

A_Y	B_Y	A_X	B_X	$A_X B_X$	$A_Y B_Y$	$A_X B_X + A_Y B_Y$ Sum($O6$)	Carry($O5$)	$O6(Hex)$	$O5(Hex)$
0	0	0	0	0	0	0	0	8	0
0	0	0	1	0	0	0	0		
0	0	1	0	0	0	0	0		
0	0	1	1	1	0	1	0		
0	1	0	0	0	0	0	0	8	0
0	1	0	1	0	0	0	0		
0	1	1	0	0	0	0	0		
0	1	1	1	1	0	1	0		
1	0	0	0	0	0	0	0	8	0
1	0	0	1	0	0	0	0		
1	0	1	0	0	0	0	0		
1	0	1	1	1	0	1	0		
1	1	0	0	0	1	1	0	7	8
1	1	0	1	0	1	1	0		
1	1	1	0	0	1	1	0		
1	1	1	1	1	1	0	1		

also be ported to the newer versions of FPGAs, such as Virtex UltraScale+. As described previously, a computational block of Fig. 3.3 is equivalent to a LUT and the one bit cell (CC) of associated carry chain in Fig. 3.6. Table 3.2 defines the *Type-A* LUT configuration for the generation and summation of partial product bits $A_Y B_Y$ and $A_X B_X$ of Fig. 3.6. The LUT initially performs the logical *AND* operation on $A_Y B_Y$ and $A_X B_X$ and then produces the O5 (generate) and O6 (propagate) signals. The values $O5 = 0x8000$ and $O6 = 0x7888$ accommodate only four input values. As discussed previously in Sect. 2.2, the INIT value for LUT6 to produce $O5 = 0x8000$ and $O6 = 0x7888$ will be $0x7888788880008000$. The INIT value for *Type-B* LUT configuration is $0xFFFFFFFF80008000$ and its configuration is shown in Table 3.3.

As shown in Fig. 3.6, three 8-bit long Processed Partial Products (PPPs) have been generated in the first stage of multiplication. Our proposed automated methodology organizes these PPPs in a single group and utilizes ternary addition for computing the final product. The ternary adder in Fig. 3.7 shows the computation of final product bits P_2–P_5 by adding three partial products in one step. The carry out of the slice is forwarded to the carry chain in the next slice for computing other product bits.

Since the proposed implementation relies on the efficient utilization of the available LUTs and carry chain in a slice, the LUTs required by an $N \times M$ multiplier can be estimated even without synthesizing the design using Eq. 3.2:

$$\text{No. of required LUTs} < \left\lfloor \frac{M}{4} \right\rfloor \times (N+4) + \left\lceil \frac{M}{2} \right\rceil \times N \qquad (3.2)$$

Table 3.3 Type-B LUT configuration

B_{M-1}	A_{N-1}	B_{M-2}	A_0	$A_0 B_{M-2}$	$A_{N-1}B_{M-1}$	$Sum(O6)$	$Carry(O5)$	$O6(Hex)$	$O5(Hex)$
0	0	0	0	0	0	0	0	0	8
0	0	0	1	0	0	0	0		
0	0	1	0	0	0	0	0		
0	0	1	1	1	0	0	1		
0	1	0	0	0	0	0	0	0	8
0	1	0	1	0	0	0	0		
0	1	1	0	0	0	0	0		
0	1	1	1	1	0	0	1		
1	0	0	0	0	0	0	0	0	8
1	0	0	1	0	0	0	0		
1	0	1	0	0	0	0	0		
1	0	1	1	1	0	0	1		
1	1	0	0	0	1	1	0	F	8
1	1	0	1	0	1	1	0		
1	1	1	0	0	1	1	0		
1	1	1	1	1	1	1	1		

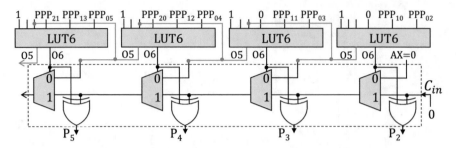

Fig. 3.7 Virtex 7 FPGA slice-based ternary adder: computation of final product bits P_2–P_5 for a 6×6 *Acc* multiplier

3.5   Motivation for Signed Multipliers

For some signed numbers-based applications, it may still be possible to implement the required hardware accelerators utilizing unsigned multiplier designs. For example, consider the fully connected artificial neuron, of a lightweight ANN, shown in Fig. 3.8a. We have quantized the trained parameters (weights and biases) of the ANN to 8-bit fixed-point numbers to implement the ANN on FPGA. These parameters are signed numbers. As shown in Fig. 3.8b, two different designs can be used for the multiplication of signed numbers. For an unsigned multiplier-based implementation of the ANN hardware, dedicated units and signed-unsigned converters are required to extract the sign bit from the operands and to compute the sign of the final product. These converters receive 2's complement numbers and produce corresponding numbers in sign-magnitude format. After multiplication in sign-magnitude format, the result is converted back to 2's complement scheme using a signed-unsigned converter. Table 3.4 shows the implementation results of the two designs for an unsigned 8×8 multiplier for Virtex-7 family FPGA using Xilinx Vivado 17.4.[2] These additional modules have increased the critical path delay of each multiplier by 2.2 ns and LUTs utilization by 22. Therefore, for the hardware implementations of applications utilizing signed numbers, it is always advantageous to have high-performance signed arithmetic units. However, most of the state-of-the-art techniques are for unsigned multipliers only. In the following sections, we present various designs of signed multipliers.

---

[2] The performance impact of signed-unsigned converters is further elaborated in Table 3.5.

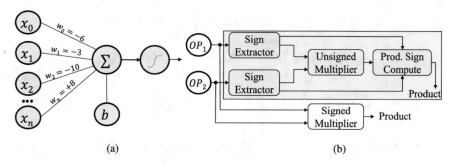

Fig. 3.8 An example of utilizing unsigned multipliers for signed numbers-based applications. (**a**) A fully connected neuron. (**b**) Signed vs. unsigned multipliers

Table 3.4 Impact of signed-unsigned converters on resource utilization and CPD

	Without sign converters		With sign converters	
Multiplier	LUTs	CPD [ns]	LUTs	CPD [ns]
8×8 [9]	51	3.86	73	6.08

3.6 Baugh-Wooley's Multiplier: *Mult-BW*

Utilizing the proposed design flow, presented in Fig. 3.2, and Baugh-Wooley's multiplication algorithm, described in Eq. 2.5, we present our novel design of accurate signed multiplier. For an $N \times M$ signed multiplier, our methodology generates only $\left\lceil \frac{M}{2} \right\rceil$ signed PPPs. This feature of our proposed implementation is similar to the commonly used radix-4 Booth's multiplication algorithm, which halves the total number of generated partial products [6] Further, Baugh-Wooley's algorithm eliminates the need for extra sign-extension bits, which help realize a resource-efficient implementation of the multiplier. Figure 3.9 presents the graphical representation of Baugh-Wooley's algorithm. As shown, the last partial product row and the most significant term in all other partial product rows are complemented. To accommodate the generation of these complemented terms, we update our proposed design flow with three new LUTs configurations. Figure 3.10 presents the new configurations of LUTs. Utilizing these configurations, Fig. 3.11 presents the *LUTs and Carry Chain Assignment* step of our proposed methodology for an $N \times M$ signed multiplier. After generating all signed PPPs, we utilize the *rearrangement and reduction of PPPs* step of our proposed methodology to compute the final product. Further, the 1's at bit locations 2^{N-1}, 2^{M-1}, and 2^{N+M-1}, as shown in Fig. 3.9, are also added during the final step of PPPs reduction.

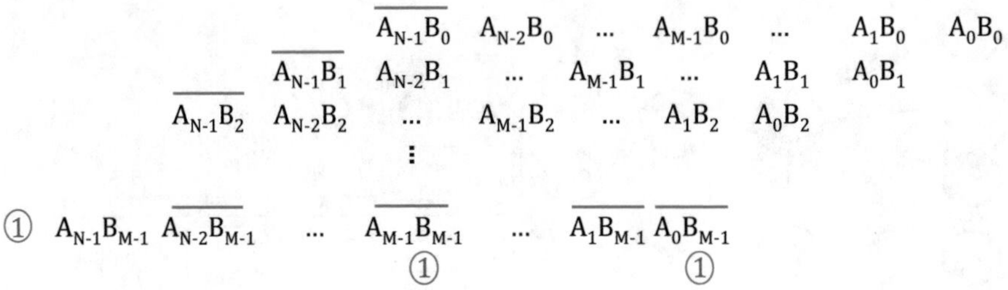

$$\begin{array}{c}
\overline{A_{N-1}B_0} \;\; A_{N-2}B_0 \;\; \cdots \;\; A_{M-1}B_0 \;\; \cdots \;\; A_1B_0 \;\; A_0B_0 \\
\overline{A_{N-1}B_1} \;\; A_{N-2}B_1 \;\; \cdots \;\; A_{M-1}B_1 \;\; \cdots \;\; A_1B_1 \;\; A_0B_1 \\
\overline{A_{N-1}B_2} \;\; A_{N-2}B_2 \;\; \cdots \;\; A_{M-1}B_2 \;\; \cdots \;\; A_1B_2 \;\; A_0B_2 \\
\vdots
\end{array}$$

① $A_{N-1}B_{M-1}$ $\overline{A_{N-2}B_{M-1}}$ $\cdots$ $\overline{A_{M-1}B_{M-1}}$ $\cdots$ $\overline{A_1B_{M-1}}$ $\overline{A_0B_{M-1}}$
                                                         ①   ①

Fig. 3.9 Baugh-Wooley's $A_{N\text{-}bit} \times B_{M\text{-}bit}$ signed multiplier design

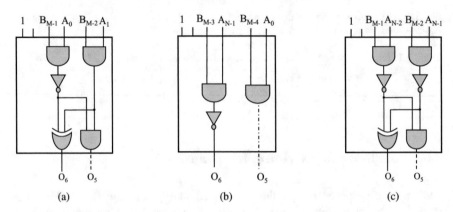

Fig. 3.10 LUTs configuration for accurate signed multiplier. (**a**) LUT of Type-C. (**b**) LUT of Type-D. (**c**) LUT of Type-E

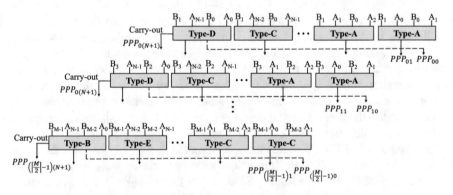

Fig. 3.11 Partial product generation for an $N \times M$ proposed accurate signed multiplier

3.7   Booth's Algorithm-Based Signed Multipliers

Utilizing the concepts of Booth's multiplication algorithm (Sect. 2.3.2) and the efficient sign extension technique (Sect. 2.3.3), we present three different designs of signed multipliers. The first implementation, referred to as *Booth-Mult*, is focused on resource optimization by combining the process of partial product generation and their accumulation. The proposed $M \times N$ *Booth-Mult* further supports the addition of an M-bit number to provide a sort of MAC functionality. We then analyze *Booth-Mult* and perform FPGA-specific resource optimizations to propose *Booth-Opt* multiplier with further reduction in the overall resources of the multiplier. The third design, denoted as *Booth-Par*, targets reducing the critical path delay of the multiplier by computing all partial products in parallel and then adding the generated partial products using multiple 4:2 compressors and a Ripple Carry Adder (RCA). The parallel generation of partial products significantly reduces the critical path delay of the multiplier.

3.7.1 Booth-Mult Design

Utilizing the 6-input LUTs and associated fast carry chains of modern FPGAs, we present an area-optimized and energy-efficient implementation of radix-4 Booth multiplier—*Booth-Mult*. The Booth's encoding scheme, shown in Table 2.1, is implemented by the LUT configuration *Type-A* shown in Fig. 3.12a. It receives six inputs, i.e., a_m, a_{m-1} (from multiplicand), b_{n+1}, b_n, b_{n-1} (from multiplier), and pp_{in} (partial product sum from previous row). Depending upon the shift flag "s," either a_m or a_{m-1} will be forwarded. Similarly, depending upon the complement flag "c," the 1's complement of a partial product can be forwarded. The third MUX, controlled by zero flag "z," can make partial product zero if the "z" flag = 1. The output of the third MUX is *XORed* with the partial sum (pp_{in}) and forwarded to

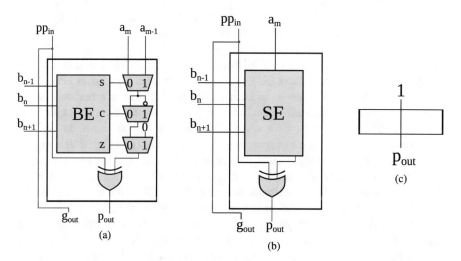

Fig. 3.12 Configuration of LUTs used in proposed methodology for $A_{M\text{-}bit} \times B_{N\text{-}bit}$ Booth's algorithm-based signed multiplier. (**a**) LUT of Type-A. (**b**) LUT of Type-B. (**c**) LUT of Type-Z

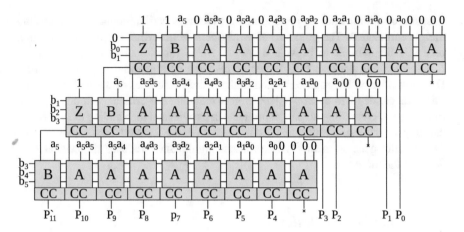

Fig. 3.13 A 6 × 6 accurate multiplier implementation (Booth-Mult)

associated carry chain as carry-propagate signal. The carry-generate signal for the carry chain is provided by the pp_{in}. The sign extension of each partial product is implemented by the LUT configurations *Type-B* and *Type-Z*, shown in Fig. 3.12b and c, respectively. LUT Type-B performs XOR operation between the SE and pp_{in} signals. The output of this operation is forwarded as propagate signal to the carry chain. The pp_{in} is also used as the carry-generate signal. LUT Type-Z represents the most significant constant "1" in a partial product row, as shown in Fig. 2.2, and is used for forwarding the sign extension information to higher-order partial product rows using sum and carry output bits of the carry chain.

Utilizing LUTs of Type-A, Type-B, and Type-Z, Fig. 3.13 shows the implementation of a 6 × 6 signed multiplier. As described in Fig. 2.2, a C_x is added at the least significant bit (LSB) position of each partial product for representing it in 2's complement format. This task of finding the correct C_x is performed by the rightmost LUT in each partial product row in Fig. 3.13. This carry will be used by the carry chain element of next LUT of Type-A. The most significant two LUTs, LUT Type-B and LUT Type-Z, in each partial product row are responsible for implementing correct sign extension. LUT Type-B computes the correct sign bit, and the LUT Type-Z is used for adding the constant "1" as shown in Fig. 2.2. However, the last partial product row does not contain a LUT of Type-Z. Due to the very regular pattern of our proposed multiplier implementation, the LUTs required for implementing an $M \times N$ multiplier can be estimated by Eq. 3.3, where "M" is the multiplicand and "N" is the multiplier. Since "N" defines the number of partial product rows in an implementation, mutual swapping of multiplicand and multiplier for, *Multiplicand* < *Multiplier*, can result in a more resource-efficient design. As shown in Fig. 3.13, the "pp_{in}" signals of LUT Type-A have been initialized to constant "0" in the first partial product row. For an $M \times N$ implementation of the proposed multiplier, an M-bit number can be further added using these "pp_{in}" signals of the first partial product row to achieve the MAC operation. Since digital signal processing applications frequently utilize MAC operations, our proposed accurate multiplier can be very useful for such applications to obtain significant area gains:

$$\text{LUTs for } M \times N \text{ multiplier } = (M+4) \times \left\lceil \frac{N}{2} \right\rceil - 1 \qquad (3.3)$$

3.7.2 Booth-Opt Design

The analysis of the implementation shown in Fig. 3.13 reveals the following observations:

- The first two LUTs in each partial product row are underutilized. The first LUT has three constant "0" inputs, and only the carry output of its associated carry chain is used. Similarly, the second LUT has also a constant "0" input. It is possible to achieve the functionalities of these two LUTs, in each partial product row, using a single modified LUT shown in Fig. 3.14a. The shaded MUXes and XOR gate perform similarly to those in LUT Type-A, defined in Fig. 3.12a. However, the non-shaded MUXes are responsible for generating and forwarding the correct carry to the next LUT Type-A in the same partial product row. The generated carry is also used for producing the least significant product bit, of the respective partial product row, using the non-shaded XOR gate.
- LUT Type-Z, in each partial product row, is used for forwarding sign extension information to other partial product rows. It forwards a constant "1" as the carry-propagate signal to the associated carry chain. As presented in Sect. 2.2, this will result in SUM $= \overline{CIN}$ and COUT $=$ CIN. The $\overline{CIN}$ is then used by a LUT Type-A in the succeeding partial product row. However, instead of using LUT Type-Z for generating $\overline{CIN}$, the LUT Type-A can be modified to invert an incoming signal internally. This results in LUT Type-A2 shown in Fig. 3.14b.

Utilizing LUTs Type-A1 and Type-A2, Fig. 3.15, shows an area-optimized implementation of a 6×6 accurate signed multiplier. Each partial product row starts with LUT Type-A1. The LUT Type-A2, in first partial product row, with inputs a_5 and constant "1" is identical to LUT Type-A with inputs a_5 and constant "0." The LUT Type-A in first partial product row is replaced with LUT Type-A2 for making the first partial product row identical to other partial product rows. A carry out of carry chain element associated with a LUT Type-B is forwarded to LUT Type-A2 and Type-B in succeeding partial product row. For this area-optimized multiplier implementation, the total number of LUTs required for implementing an $M \times N$ multiplier is represented in Eq. 3.4:

$$\text{LUTs for } M \times N \text{ multiplier } = (M+2) \times \left\lceil \frac{N}{2} \right\rceil \qquad (3.4)$$

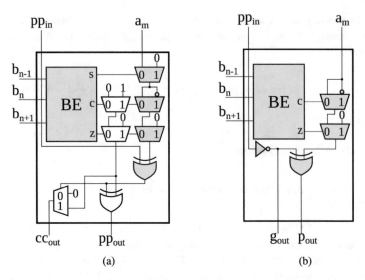

Fig. 3.14 Modified configuration of LUTs for implementing *Booth-Opt* multiplier. (**a**) LUT Type-A1. (**b**) LUT Type-A2

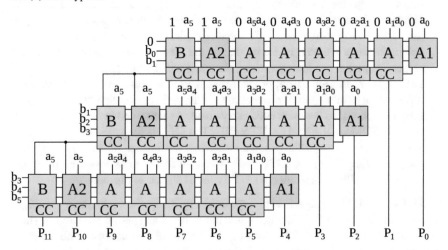

Fig. 3.15 A 6×6 area-optimized accurate multiplier (Booth-Opt)

3.7.3 Booth-Par Design

Booth-Par design eliminates the need for sequential computation of the partial products and generates all Booth-encoded partial products in parallel; that significantly reduces the overall critical path delay of the multiplier. The proposed partial product encoding technique further reduces the length of the carry chain in each partial product to reduce the critical path of the multiplier.

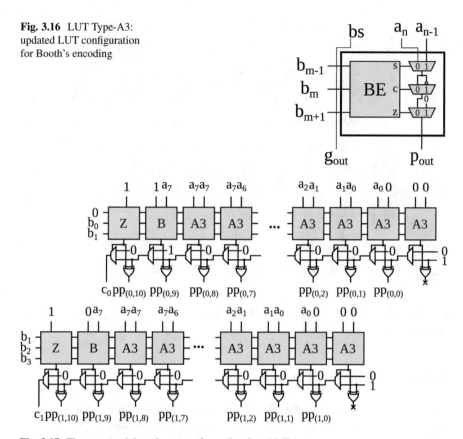

Fig. 3.16 LUT Type-A3: updated LUT configuration for Booth's encoding

Fig. 3.17 First two partial product rows for an 8×8 multiplier

Figure 3.16 shows the updated configurations of the 6-input LUT (Type-A3) used to implement the Booth's encoding (BE), described in Table 2.1. It receives five inputs, i.e., a_n and a_{n-1} (from multiplicand) and b_{m+1}, b_m, and b_{m-1} (from multiplier). The LUT internally implements three MUXes. Based on the value of BE, the first MUX (controlled by s signal) decides whether a_n or a_{n-1} should be forwarded for partial product generation. The second MUX, controlled by c signal, manages the inversion of the output of the first MUX. Finally, the third MUX can make the partial product zero depending upon the value of the z signal. This information is forwarded to the associated carry chain as carry-propagate signal "p_{out}." The carry-generate signal "g_{out}" is provided by the external bypass signal bs.

Utilizing LUTs of Type-A3, Type-B, and Type-Z, Fig. 3.17 shows the first two rows of partial products for an 8×8 *Booth-Par* multiplier. As mentioned previously, the rightmost LUT of Type-A in each partial product row is used for computing the required input carry. This input carry is applied for representing a partial product in 2's complement format. As described in Sect. 2.3, a radix-4 Booth's encoding halves the total number of generated partial products; therefore, an 8×8 *Booth-*

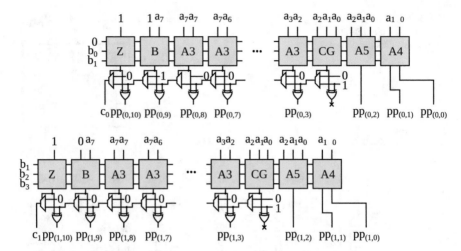

Fig. 3.18 First two partial product rows for an optimized 8×8 multiplier

Par multiplier generates four partial product rows. The last partial product row does not require a LUT of Type-Z.

3.7.3.1   Optimizing Critical Path Delay

For an $N \times M$ multiplier, the length of the carry chain in each partial product row is $N + 4$ bits. To improve the critical path delay of the multiplier, the length of the carry chain can be reduced to $N + 1$ bits. A critical path delay-optimized implementation of our novel multiplier is shown in Fig. 3.18. The partial product terms $pp_{(x, 0)}$ and $pp_{(x, 1)}$, in each partial product row, require one and two bits of multiplicand, respectively. These two partial product terms can be implemented by one single 6-input LUT "A4." Similarly, $pp_{(x, 2)}$, in each partial product row, can be independently implemented using another 6-input LUT "A5." A separate 6-input LUT, "CG," can be used to compute the correct input carry for each partial product row. Figure 3.19 shows the internal configurations of LUT Type-A4, Type-A5, and Type-CG, respectively. LUT Type-A5 and Type-CG only differ in the output signal $pp_{(x, 2)}$ and cg_{out}. LUT Type-A5 utilizes $pp_{(x, 2)}$ signal solely, whereas LUT Type-CG uses cg_{out} signal exclusively. For an $N \times M$ multiplier, the number of LUTs required to generate partial products is $(N + 3) \times \left\lceil \frac{M}{2} \right\rceil - 1$.

3.7.3.2   Accumulation of Generated Partial Products

For the reduction of generated partial products to compute the final product, binary adders, ternary adders, and 4:2 compressors [21] can be utilized. A 4:2 compressor

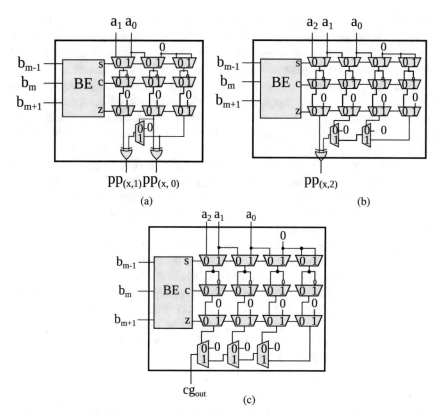

Fig. 3.19 Configuration of LUTs Type-A4, Type-A5, and Type-CG for implementing *Booth-Par*. (**a**) LUT Type-A4. (**b**) LUT Type-A5. (**c**) LUT Type-A5

is capable of reducing four partial product rows to two output rows. During our experiments, we observe that the deployment of ternary adders might reduce the overall resource utilization. However, they have higher critical path delays than the binary adders. Therefore, in our work, the 4:2 compressors and binary adders are used for the reduction of the generated partial products. We have used the 6-input LUTs and the associated carry chains to implement them.

3.8 Constant Multipliers

For many applications such as RL, one of the parameters in multiplication operations remains constant. For such multiplication operations, the utilization of a constant multiplier can offer significant reductions in the overall resource utilization, critical path delay, and energy consumption. One of the inputs to a constant multiplier is a fixed constant number, and the other input is a variable. FPGA synthesis tools, such as Xilinx Vivado, provide both DSP-based and LUT-based constant multiplier implementations. However, in this book, we have utilized the 6-input LUTs and the associated carry chains to implement constant multipliers with better resource optimization than the multiplier IP provided by Vivado.

Algorithm 1 presents our proposed technique for efficiently implementing constant multipliers. Our proposed method efficiently utilizes binary and ternary adders[3] for computing the product. As described in Algorithm 1, to multiply an M-bit variable V with an N-bit constant C, we first identify the total number of $1's$, referred to as C_1, in the binary representation of C (line 1). To reduce the total number of shift and add operations in the multiplication, we compare C_1 with $N - C\_1$ (line 2). For cases where $N - C\_1$ is smaller than C_1, we try to find the product $C \times V$ by shifting V left N times and then subtracting X-times shifted version of V from it (lines 3–6). For example, the 8-bit representation of constant 254 is '$0b11111110$'. The C_1 and C_0 are 7 and 1, respectively. In this case, we compute the parameter $X = 2^8 - 254 = 2$. Our proposed implementation initially computes $2^8 \times V$ (shift variable "V" left by 8) and then subtracts $2 \times V$ (shift variable "V" left by 1) from it. However, for all cases where C_1 is smaller than $N - C\_1$, we compute the product using the left shift and add operations (line 11). Similarly, for cases where the number of levels to compute product using ternary adders is smaller than the number of levels using subtraction (line 5), we compute product using left shift and add operations (line 8). For example, for constant $C = 249$ ('$0b11111001$'), parameter "X" is 7('$0b00000111$') and the number of 1's in the binary representation of "X" ($X\_1 = 3$) is not less than the $\left\lceil \frac{C\_1}{3} \right\rceil$; therefore, the constant multiplier is instead implemented using binary *shift* and *add* operations. Moreover, the constant multipliers can considerably reduce the overall resource utilization when the constant parameters have values that are a direct power of 2. For such values of the constant parameters, the multiplication operations can be performed by a single shift operation.

3.9 Results and Discussion

3.9.1 Experimental Setup and Tool Flow

All presented multipliers have been implemented in VHDL and synthesized for the 7VX330T device of Virtex-7 family FPGA using Xilinx Vivado 17.4. For Power-Delay Product (PDP) calculations, Vivado Simulator and Power Analyzer tools have been used. The proposed methodology implements each design multiple times with a different critical path constraint in each iteration to produce precise area (LUTs), Critical Path Delay (CPD), and dynamic power consumption values. In

---

[3] A ternary adder using 6-input LUTs and associated carry chain is presented in Fig. 3.7.

Algorithm 1: Constant multiplier using shift operation

Require: An N-bit positive constant 'C' and an M-bit variable 'V'

1: Identify the total number of 1's (C_1) and 0's (C_0) in the binary representation of C
2: **if** C_0 < C_1 **then**
3:     Compute X: $X = 2^N - C$
4:     Identify the total number of 1's (X_1) in the binary representation of X
5:     **if** $X\_1 < \left\lceil \frac{C\_1}{3} \right\rceil$ **then**
6:         Compute C × V = 2^N × V − X × V
7:     **else**
8:         Compute C × V using C_1 times shift left and utilizing binary and ternary adders
9:     **end if**
10: **else**
11:     Compute C × V using C_1 times shift left and utilizing binary and ternary adders
12: **end if**

---

each implementation-iteration, our automated tool flow adjusts the new critical path constraint according to the critical path-slack obtained from the previous iteration. The total number of implementation-iterations, performed by our tool flow to provide the final CPD and resource utilization information of a design, is adjustable. For this chapter, we have kept the maximum number of implementation-iterations at *10*. The final achieved minimum CPD of a design is then used for computing its dynamic power values using the Vivado Simulator and Power Analyzer tools.

3.9.2 Performance Comparison of the Proposed Accurate Unsigned Multiplier Acc

To evaluate the efficacy of the proposed accurate unsigned multiplier implementation, we compare it to the existing standard multipliers, such as Xilinx LogiCORE Multiplier *IP* (area and speed optimized) [3], Booth Multipliers (*S3*) [9], Xilinx's default K-Map solved optimized multiplier (*S4*) [22], Wallace Tree (*S5*) [17], and Dadda (*S6*) [18] multipliers. Figure 3.20 compares the LUT utilization and CPD requirements of different unsigned accurate multipliers for different bit widths. The proposed *Acc* multiplier always lies on the area-delay Pareto fronts of different sizes of multipliers. The *S5* and *S6* implementations consume a large number of LUTs, as shown in Fig. 3.20b; therefore, they have not been considered for designing higher-order multipliers. For example, compared to the proposed 8 × 8 *Acc* multiplier, *S5* and *S6* require 82.7% and 201.9% more LUTs for implementation. Even though *S3* occupies fewer LUTs for higher-order multipliers, the sequential generation-addition of partial products in *S3* results in higher critical path delays. For example, compared to the proposed 16 × 16 *Acc* multiplier, *S3* has a 45.5% higher CPD. The dynamic power consumption of all of the above implementations

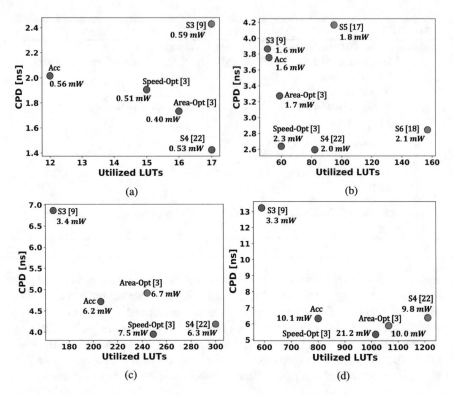

Fig. 3.20 Area and critical path delay results of different accurate unsigned multipliers. (**a**) 4 × 4. (**b**) 8 × 8. (**c**) 16 × 16. (**d**) 32 × 32

has also been shown along with each implementation. Compared to Vivado area- and speed-optimized multiplier IPs, our proposed *Acc* multiplier provides better resource utilization and power consumption across different sizes of multipliers. For example, compared to the 16 × 16 area-optimized multiplier IP, the proposed *Acc* multiplier provides a 15.5 and 7.4% reduction in the total utilized LUTs and power consumption, respectively.

3.9.3   Performance Comparison of the Proposed Accurate Signed Multiplier with the State-of-the-Art Accurate Multipliers

We compare our proposed accurate signed multipliers (*Mult-BW*, *Booth-Opt*, and *Booth-Par*) with Booth's multiplication algorithm-based (S3) [9] , and Vivado area- and speed-optimized multiplier IPs [3]. The works in [7] and [8] have also used 6-input LUTs to implement signed multipliers. However, the implementations in

[7] and [8] have used a target design period of 1 ns, and the corresponding results show that none of the implementations meets the target design period. Therefore, in this book, we do not compare our implementation results with those mentioned in these articles. Table 3.5 shows the comparison of the resource utilization, critical path delay, and energy consumption (Power-Delay Product (PDP)) of the proposed signed multipliers with different state-of-the-art accurate multipliers. We have also compared our proposed multipliers with the signed version of the *S3* multiplier. For this purpose, we have used signed-unsigned converters. In the table, the results for *S3* multiplier are inclusive of the signed-unsigned converters.

As shown by the results, the proposed designs provide a fair trade-off between the total number of utilized LUTs and the corresponding achieved CPD and PDP. The *Mult-BW* and the *Booth-Par* designs generate all partial products simultaneously and then deploy different techniques for the summation of the generated partial products. The parallel generation of all partial products followed by a separate partial products reduction tree significantly reduces the multiplier's critical path delay. For example, *Booth-Par* utilizes 4:2 compressors and binary adders to reduce partial products, and the implementation offers the minimum CPD across different sizes of the multiplier. The *Mult-BW* design utilizes ternary and binary adders for the addition of generated partial products, as discussed in Sect. 3.6. The utilization of ternary adders enables resource-efficient implementations. However, the dependency of every element of the carry chain on the carry-generate signal from its preceding cell, as shown in Fig. 3.7, diminishes the performance of a ternary adder. For example, the implementation of an 8-bit ternary adder (three operands) on Virtex-7 FPGA using Vivado has a 37% higher CPD compared to an 8-bit binary adder (two operands). However, the separation between partial product generation and their reduction results in utilizing more LUTs. The *Booth-Opt* design fuses these two operations and utilizes fewer LUTs than the *Booth-Par* and *Mult-BW* designs across different sizes of multipliers. However, the sequential generation of the *Booth-Opt* partial products results in increasing the CPD of the implementation. For smaller multipliers, such as 4×4 and 8×8, the *Mult-BW* and *Booth-Opt* have comparable CPD. However, for larger multipliers, such as 16×16 and 32×32, *Mult-BW* offers lower CPD than *Booth-Opt* implementation. The reasons for comparable CPDs of the *Mult-BW* for smaller multipliers are the higher latency of the ternary adders and the addition of supplementary $1's$ in the partial product reduction tree, as shown in Fig. 3.9. For smaller multipliers, these result in diminishing the advantages of the parallel generation of partial products.

Compared to Vivado area- and speed-optimized multiplier IPs, our proposed designs always utilize fewer LUTs across various sizes of multipliers. For example, compared to 8×8 area-optimized multiplier IP, *Mult-BW*, *Booth-Opt*, and *Booth-Par* offer 38.6%, 54.5%, and 25% reduction in the total utilized LUTs, respectively. Overall, the *Booth-Par* implementation provides lower CPD when compared with other designs. For example, compared to Vivado speed-optimized 8×8 multiplier IP, *Booth-Par* offers a 20.9% reduction in the CPD. For larger multipliers, such as 16×16 and 32×32, Vivado speed-optimized IP has a slightly reduced CPD than the *Booth-Par* design at the cost of a significant increase in the total utilized LUTs. The

Table 3.5 Implementation results of different multipliers. The *S3* multipliers are implemented with the signed-unsigned converters. The CPD and PDP are in ns and pJ, respectively. The results with shading are the lowest in their respective column

Design	4 × 4			8 × 8			16 × 16			32 × 32			Average performance
	LUTs	CPD	PDP	LUTs	CPD	PDP	LUTs	CPD	PDP	LUTs	CPD	PDP	
Mult-BW	14	2.59	1.312	54	4.37	7.26	208	5.28	31.14	803	7.35	63.75	0.49
Booth-Opt	12	2.15	1.09	40	4.25	5.14	144	7.64	21.15	544	15.18	44.56	0.34
Booth-Par	18	1.65	1.13	66	2.80	6.06	243	4.48	26.52	928	6.08	57.50	0.38
S3 [9]	24	3.84	2.49	73	6.08	9.70	217	9.52	32.94	700	16.25	100.42	1.37
IP speed [3]	18	2.14	1.06	74	3.54	5.73	286	4.27	33.91	1103	5.81	104.46	0.60
IP area [3]	30	2.91	2.25	88	3.45	9.07	326	5.04	35.25	1102	6.79	95.21	1

Booth-Opt design offers better energy efficiency than other designs by having lower PDP values. For example, compared to Vivado area-optimized 8×8 multiplier IP, *Booth-Opt* reduces the energy consumption by 43.3%.

We have also compared our proposed multipliers with the signed version of the *S3* multiplier using signed-unsigned converters. Our proposed multipliers provide better resource utilization, lower CPD, and higher energy efficiency than the signed *S3* multiplier. For example, compared to our proposed 8×8 *Booth-Opt* multiplier, the *S3* multiplier has an 82.5%, 43.0%, and 88.5% increase in the total utilized LUTs, CPD, and PDP, respectively. Compared to performance metrics of the unsigned *S3* multiplier shown in Fig. 3.20, the utilization of signed-unsigned converters has significantly degraded the overall performance of the multiplier. For example, the employment of the converters for an 8×8 *S3* multiplier results in an increase of 43.1%, 57.1%, and 54.8% in the total utilized LUTs, CPD, and PDP, respectively. Therefore, it is always beneficial to have optimized implementations of unsigned and signed multipliers as motivated in Sect. 3.5.

To further elaborate on the efficacy of our proposed implementation, Table 3.5 shows the averages of the products of the normalized values of LUTs utilization, CPD, and PDP (Average [Norm. LUTs × Norm. CPD × Norm. PDP]) across different sizes of multipliers. All individual performance metrics of each multiplier have been normalized to the corresponding performance metrics of Vivado area-optimized multiplier IP. A smaller average value of the metric presents an implementation with a better performance. Our proposed multipliers outperform state-of-the-art implementations in the overall score.

We also compare our proposed accurate designs with DSP blocks-based multipliers. These results are presented in Table 3.6. To provide a thorough comparison, we have explored the various synthesis optimization strategies provided by the Xilinx Vivado synthesis tool for DSP blocks-based multipliers, such as area/speed optimization and unsigned/signed operations. However, the performance metrics of DSP blocks-based multipliers do not have a significant difference between unsigned and signed numbers-based operations; therefore, we have shown the results for only signed numbers-based DSP blocks. Compared to the proposed accurate multiplier implementations, the DSP blocks-based multipliers have higher CPD and PDP values for lower bit-widths multipliers, such as 4×4 and 8×8. For example, the proposed 4×4 *Mult-BW* multiplier offers a 23% and 70.6% reduction in CPD and PDP values, respectively, when compared with the area-optimized 4×4 DSP blocks-based multiplier. The DSP blocks-based multiplier's degraded performance is because the DSP48E1 slice in 7 series FPGAs (used for the computation of experimental data in Table 3.6) hosts a 25×18 multiplier and is not optimized for smaller multipliers [1]. According to the design recommendations of Xilinx Vivado [1], LUTs-based soft multipliers should be used for implementing lower bit-widths multipliers. Our proposed multipliers provide a feasible trade-off between performance and resource utilization for such scenarios. For higher-order multipliers, such as 16×16, the DSP blocks-based multipliers provide reduced CPD and PDP values than our proposed multipliers. However, the proposed *Booth-Par* has a lower CPD than the DSP blocks-based multipliers for 32×32 multiplication.

Table 3.6 Performance comparison of proposed multipliers with DSP blocks-based multipliers. The CPD and PDP are in ns and pJ, respectively

Design	4 × 4				8 × 8				16 × 16				32 × 32			
	LUTs	DSPs	CPD	PDP	LUTs	DSPs	CPD	PDP	LUTs	DSPs	CPD	PDP	LUTs	DSPs	CPD	PDP
Acc (unsigned)	12	0	2.0	1.1	52	0	3.8	6.1	206	0	4.7	29.6	800	0	6.3	64.1
Mult-BW (signed)	14	0	2.6	1.3	54	0	4.4	7.3	208	0	5.3	31.1	803	0	7.4	63.8
Booth-Opt (signed)	12	0	2.2	1.1	40	0	4.3	5.1	144	0	7.6	21.2	544	0	15.2	44.6
Booth-Par (signed)	18	0	1.7	1.1	66	0	2.8	6.1	243	0	4.5	26.5	928	0	6.1	57.5
Vivado IP area opt. [1]	0	1	3.4	4.5	0	1	3.6	5.5	0	1	3.6	5.9	611	1	7.3	56.4
Vivado IP speed opt. [1]	0	1	3.4	4.5	0	1	3.5	5.5	0	1	3.7	6.0	0	4	6.9	22.4

For 32×32 multiplication, the synthesis tool utilizes more than one DSP block, and the resulting interconnection of the DSP blocks results in degrading the overall performance of the multiplier. Further, compared to the proposed 32×32 multipliers, the area-optimized DSP blocks-based multiplier utilizes one DSP slice and 611 LUTs. The corresponding speed-optimized 32×32 IP utilizes 4 DSP slices.

The utilization of DSP blocks along with a large number of LUTs for DSP blocks-based multipliers call for the orthogonal approach of defining resource-efficient soft multiplier architectures for multiplier-intensive applications, such as Artificial Neural Networks (ANNs), implemented on a small FPGA. Towards this end, we experimented on a small Multi-layer Perceptron (MLP) to classify the MNIST dataset [23]. The inference accuracy of the dataset using the single-precision floating-point number is 97%. The corresponding inference accuracy using 8-bit fixed-point quantization is 96.6%, resulting in an insignificant drop in output accuracy. To evaluate the performance metrics of the quantized MLP implementation on FPGA, we implemented a single layer of the MLP on Xilinx Zynq UltraScale+ MPSoC[4] (xczu3eg-sbva484-1-e device). The implementation consists of instantiating 20 neurons with 128 input activations. The experiment results using 8×8 DSP blocks-based multipliers and the proposed *Mult-BW* multiplier are presented in Table 3.7. The DSP blocks-based design offers a lower critical path delay than the *Mult-BW* multiplier-based implementation. However, the DSP blocks-based implementation requires overall more resources than the *Mult-BW* multiplier-based implementation. As shown, the DSP blocks-based implementation utilizes 54.5 and 34.6% of the total available DSP blocks and LUTs, respectively, on the FPGA. However, the corresponding implementation using our proposed *Mult-BW* multiplier requires only 46.4% of the total available LUTs. However, the resource efficiency offered by a particular type of operator can be utilized to instantiate more instances of the operator to increase the overall performance of an implementation. For example, Table 3.7 shows the maximum theoretical performance, Giga Operations per Second (GOPS), of the two implementations of MLP by instantiating the maximum number of DSP blocks-based and proposed *Mult-BW* multipliers. The resource efficiency of *Mult-BW* allows instantiating more numbers of multipliers and providing higher performance than

Table 3.7 MLP implementation results on FPGA

Design	LUTs utilization [%]	DSPs utilization [%]	CPD [ns]	Max. performance[a] [GOPS]
DSP Blocks	34.6	54.5	53.4	4.1
Mult-BW	46.4	0.0	57.2	4.5

[a]We report the maximum theoretical performance for both implementations by exhausting the corresponding DSP blocks and LUTs

---

[4] UltraScale+ architecture has DSP blocks with 27×18 multipliers.

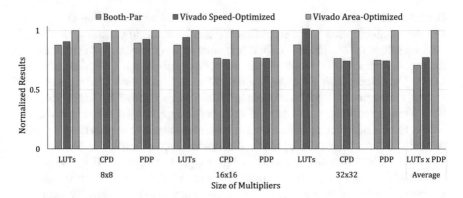

Fig. 3.21 The results of the neural network use case. The LUT resources, CPD, and PDP obtained for designs with different multiplier are normalized to Vivado area IP

the DSP blocks-based design for the benchmark application. It is obvious to use both DSP blocks and soft multipliers to realize high-performance accelerators on resource-constrained FPGAs. For example, the maximum theoretical performance of the MLP experiment by initially exhausting the DSP blocks-based multipliers and then utilizing the remaining LUTs to instantiate proposed *Mult-BW* multipliers is 5.5 GOPS.

To further evaluate the efficacy of the proposed multipliers, we have conducted the same experiment (ANN implementation) using *Booth-Par* multiplier and Vivado area- and speed-optimized soft (LUTs-based) multiplier IPs. As presented in Table 3.5, the *Booth-Par* design, among the proposed multipliers, provides the minimum CPD at the cost of more LUTs for various sizes of multipliers. However, it provides better overall performance than Vivado multiplier IPs. The same conclusion about the performance of *Booth-Par* can be observed in the ANN implementation results, shown in Fig. 3.21. First, we implement the network with Vivado speed-optimized multiplier with as many neurons as possible in three different input sizes, 8×8, 16×16, and 32×32. The timing constraint is kept at 4 ns. After that, the same setups are applied for the Vivado area-optimized multiplier and *Booth-Par* multipliers. The resulting LUT utilization, CPD, and PDP of each design are normalized to Vivado area-optimized IPs. In the combined LUT $\times$ PDP average across all input sizes, *Booth-Par* offers the best results. Our proposed multiplier is 8.4 and 29.4% better than Vivado speed- and area-optimized IPs, respectively. In comparison with Vivado speed-optimized IP, *Booth-Par* is comparable in PDP but requires an average of 8% less number of LUTs.

3.9.4 Performance Comparison of the Proposed Constant Multiplier with the State-of-the-Art Accurate Multipliers

We have used the design of our constant multipliers in the accelerator design for the Reinforcement Learning (RL) application. RL involves allowing an agent to learn from its interactions with the environment. In this technique, the agent is allowed to take certain actions, and the long-term effectiveness of each action is estimated from the returns obtained by the agent subsequent to that action till it reaches the goal. This way, the agent can perform model-free learning by allowing the machine to learn dynamically from experience rather than statically from a priori data. In the RL technique, a matrix Q stores each state-action pair's value function, and a matrix R stores the rewards associated with each state-action pair. A randomly initialized Q and an arbitrarily chosen *state* are used to start the learning process. At each iteration, based on the chosen action, the next state is selected, and the corresponding reward is used to update the Q matrix, as shown in Eq. 3.5. In this equation, $\alpha \in [0, 1]$ denotes the learning rate, and $\gamma \in [0, 1]$ represents the discounting factor. The Q-value of each state-action pair (s,a) indicates the average long-term benefit of taking action a from state s while aiming for the goal state. As the design-time parameters α and γ remain constant for a design, they can be used to implement Eq. 3.5 using constant multipliers.:

$$Q_{(s,a)} = (1 - \alpha) \times Q_{(s,a)} + \alpha \times (R_{(s,a)} + \gamma \times Q_{(s+1,a)_{\max}}) \qquad (3.5)$$

We have used our proposed technique of constant multipliers for implementing state-update equation for different values of α and γ. To show the efficacy of our implementation, we have compared our technique with the state-update equations using a generic variable multiplier, Vivado constant multiplier (Const. Mult), Vivado constant multiplier IP, and a state-of-the-art constant multiplier referred to as *FloPoCo* [24]. For the Vivado constant multiplier, the constant values of the parameters have been fixed in the Hardware Description Language (HDL) code of the multiplier. This design allows the synthesis tool to optimize the multiplier implementation with respect to the rest of the hardware. The *FloPoCo* design is also based on the utilization of *bit shifts* and *addition/subtraction* operation to implement a constant multiplier. However, as described previously in Sect. 3.8, we use 6-input LUT-level optimizations to implement our proposed multiplier. Figure 3.22 presents the total number of utilized LUTs and PDP of different implementations for three different random values of α and γ. These results have been normalized to the results obtained by implementing the state-update equation using a variable multiplier. The precisions of $Q_{(s,a)}$, $R_{(s,a)}$, and $Q_{(s+1,a)_{\max}}$ are kept at 16-bit (10.6 fixed-point format) for these experiments. Further, to represent the fractional values of α and γ in the range of [0, 1], we have used a 1.10 fixed-point format (1 bit reserved for integer and 10 bits for denoting the precision). For example, $\alpha = \gamma = 0.875$ will be represented by constant $0.875 \times 2^{10} = 896$. All four designs ("Ours," "Const. Mult," "Vivado IP," and "FloPoCo") offer reductions in the total utilized

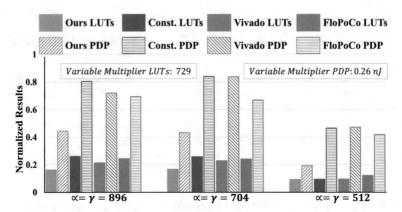

Fig. 3.22 Implementation results of constant multipliers-based state-update equation. Values are normalized to the corresponding results of state-update equation using variable multiplier

LUTs and energy consumption. Further, for all values of α and γ, our proposed constant multiplier implementation-based designs provide significant reductions in the utilized LUTs and PDP for the state-update equation. For example, compared to the Vivado IP-based implementation for $\alpha = \gamma = 704$, our proposed technique offers ~27% and ~48% reductions in the total utilized LUTs and PDP, respectively. Though the results are shown for 16-bit precision of variables and three different values of α and γ, we have observed similar reductions in the resources and energy consumption of the state-update equation for other values of α, γ, and precision of variables. The results also show a notable decrease in the LUTs utilization and energy consumption of the state-update equation for design parameters α and γ having values that are a direct power of 2.

3.10   Conclusion

In this chapter, we have explored various multiplication algorithms to provide area-optimized, high-performance softcore accurate unsigned/signed multipliers. Our proposed implementations are based on the efficient utilization of the 6-input LUTs and the associated carry chains of modern FPGAs. The LUT-level optimization and implementation of the proposed multipliers allow the accurate estimation of the total utilized LUTs and carry chains even before synthesizing the multiplier. This information is particularly beneficial in deciding the overall implementation details of an accelerator for an application. For example, given the total number of available LUTs for an FPGA, a designer can predict the total possible number of instantiated neurons for implementing an ANN. Our proposed designs provide a fair trade-off between the total utilized LUTs, Critical Path Delay, and energy consumption of the multiplier. For example, compared to the Vivado area-optimized multiplier IP, our proposed unsigned and signed multipliers provide up to 25% and 60% reduction in the overall utilized LUTs, respectively. Similarly, compared to the Vivado area-optimized multiplier IP, our proposed signed designs provide on average 66% better performance for different multiplier sizes by considering all performance metrics (LUTs $\times$ CPD $\times$ PDP). Our proposed designs also provide a better trade-off between resource utilization, CPD, and energy consumption than DSP blocks-based multipliers. For example, compared to a single DSP blocks-based 8×8 signed multiplier, our proposed *Booth-Par* utilizes 66 LUTs for the implementation and provides a 21.5% reduction in the CPD of the multiplier. We have also evaluated the efficacy of our proposed multipliers in different implementations of lightweight ANNs. Most of these designs will be used in the subsequent chapters to propose approximate multiplier designs with different accuracy-performance trade-offs.

References

1. Xilinx 7 Series DSP48E1 Slice (2018). https://www.xilinx.com/support/documentation/user_guides/ug4797SeriesDSP48E1.pdf
2. Intel^6 Stratix^6 10 Variable Precision DSP Blocks User Guide (2020). https://www.intel.com/content/dam/www/programmable/us/en/pdfs/literature/hb/stratix-10/ug-s10-dsp.pdf
3. Xilinx LogiCORE IP v12.0 (2015). https://www.xilinx.com/support/documentation/ipdocumentation/multgen/v120/pg108-multgen.pdf
4. Intel. *Integer Arithmetic IP Cores User Guide* (2020). https://www.altera.com/en_US/pdfs/literature/ug/uglpmaltmfug.pdf
5. C.R. Baugh, B.A. Wooley, A two's complement parallel array multiplication algorithm. IEEE Trans. Comput. **C-22**(12), 1045–1047 (1973)
6. A.D. Booth, A signed binary multiplication technique.Q. J. Mech. Appl. Math. **4**(2), 236–240 (1951)
7. E.G. Walters, Array multipliers for high throughput in xilinx FPGAs with 6-input LUTs. Computers **5**(4), 20 (2016)
8. E.G. Walters, Partial-product generation and addition for multiplication in FPGAs with 6-input LUTs, in *2014 48th Asilomar Conference on Signals, Systems and Computers* (IEEE, Piscataway, 2014), pp. 1247–1251
9. M. Kumm, S. Abbas, P. Zipf, An efficient softcore multiplier architecture for Xilinx FPGAs, in *2015 IEEE 22nd Symposium on Computer Arithmetic* (IEEE, Piscataway, 2015), pp. 18–25
10. M. Kumm, J. Kappauf, M. Istoan, P. Zipf, Resource optimal design of large multipliers for FPGAs, in *2017 IEEE 24th Symposium on Computer Arithmetic (ARITH)* (IEEE, Piscataway, 2017), pp. 131–138
11. J. Faraone, M. Kumm, M. Hardieck, P. Zipf, X. Liu, D. Boland, P.H.W. Leong, Addnet: Deep neural networks using fpga-optimized multipliers. IEEE Trans. Very Large Scale Integr. Syst. **28**(1), 115–128 (2019)
12. H. Parandeh-Afshar, P. Ienne, Measuring and reducing the performance gap between embedded and soft multipliers on FPGAs, in *2011 21st International Conference on Field Programmable Logic and Applications* (IEEE, Piscataway, 2011), pp. 225–231
13. Xilinx. 7 Series FPGAs Configurable Logic Block (2016). https://www.xilinx.com/support/documentation/user_guides/ug4747SeriesCLB.pdf
14. H. Parandeh-Afshar, P. Brisk, P. Ienne, Exploiting fast carrychains of FPGAs for designing compressor trees, in *2009 International Conference on Field Programmable Logic and Applications* (IEEE, Piscataway, 2009), pp. 242–249
15. M. Langhammer, G. Baeckler, High density and performance multiplication for FPGA, in *2018 IEEE 25th Symposium on Computer Arithmetic (ARITH)* (IEEE, Piscataway, 2018), pp. 5–12

16. B. Parhami, *Computer Arithmetic*, vol. 20. 00 (Oxford University Press, Oxford, 2010)
17. C.S. Wallace, A suggestion for a fast multiplier.IEEE Trans. Electron. Comput. **1**, 14–17 (1964)
18. L. Dadda, Some schemes for parallel multipliers. Alta Frequenza **34**, 349–356 (1965)
19. B. Millar, P.E. Madrid, E.E. Swartzlander, A fast hybrid multiplier combining Booth and Wallace/Dadda algorithms, in *[1992] Proceedings of the 35th Midwest Symposium on Circuits and Systems* (IEEE, Piscataway, 1992), pp. 158–165
20. J.M. Simkins, B.D. Philofsky, Structures and methods for implementing ternary adders/subtractors in programmable logic devices. US Patent 7,274,211, 2007
21. M. Kumm, P. Zipf, Efficient high speed compression trees on Xilinx FPGAs, in *MBMV* (2014), pp. 171–182
22. Xilinx. *Vivado Design Suite User Guide* (2017). https://www.xilinx.com/support/documentation/swmanuals/xilinx20174/ug910-vivado-getting-started.pdf
23. MNIST-cnn (2016). https://github.com/integeruser/MNIST-cnn
24. M. Kumm, O. Gustafsson, M. Garrido, P. Zipf, Optimal single constant multiplication using ternary adders. IEEE Trans. Circuits Syst. II Express Briefs **65**(7), 928–932 (2016)

Chapter 4
Approximate Multipliers

4.1 Introduction

Multiplication is one of the most extensively used arithmetic operations in a wide range of applications. Error-resilient applications such as Deep Neural Networks have millions of MAC operations. For example, deep residual learning (ResNet-152) [1] has 11.3 billion MAC operations per forward pass for the processing of a single image. Therefore, for these applications, approximate multipliers can be utilized for obtaining area-optimized, low-latency, and energy-efficient implementations. Previous works have proposed different approximate multiplier designs utilizing the inherent error resilience of such error-tolerant applications. However, as discussed in Chap. 1, the approximation techniques presented in most of these works, such as [2, 3], and [4], focus on ASIC-based systems and ignore the architectural specifications of FPGAs; therefore, these techniques are less effective in gaining ASIC-like energy, performance, and area gains when used for FPGA-based systems. Towards this end, this chapter presents various novel approximate unsigned/signed approximate multipliers optimized for FPGA-based systems. For most of the designs, we have used the modular design approach. In this approach, multiple smaller sub-multipliers are combined together to compute the final product. The modular design methodology permits a more extensive design space by enabling the possible exploration of different approximation strategies for each sub-multiplier and the final reduction tree (used for adding the results of sub-multipliers). Many of these designs are based on the accurate multiplier designs presented in Chap. 3. The following are the key contributions of this chapter.

© The Author(s), under exclusive license to Springer Nature Switzerland AG 2023
S. Ullah, A. Kumar, *Approximate Arithmetic Circuit Architectures for FPGA-based Systems*, https://doi.org/10.1007/978-3-031-21294-9_4

Contributions

- *Approximate* 4 × 4 *unsigned multiplier (Approx-1)*: *Approx-1* design is based on the approximate addition of the accurate Processed Partial Products (PPPs) of the *Acc* multiplier described in Chap. 3. The utilization of approximate addition provides an opportunity to implement approximate multipliers with different accuracy and performance trade-offs. Towards this end, we propose an approximate binary adder for the addition of accurate PPPs. The proposed adder eliminates the propagation of carries and adds all the terms of two PPPs to implement an approximate 4 × 4 multiplier.
- *Approximate* 4 × 4 *unsigned multiplier (Approx-2):* We present a highly accurate, low-latency, and resource-efficient design of an unsigned approximate 4 × 4 multiplier. The proposed multiplier utilizes a novel approximate 4 × 2 multiplier as a building block and utilizes carry chain-based approximate addition of two 4 × 2 multipliers to implement the 4 × 4 multiplier.
- *Approximate* 4 × 4 *unsigned multiplier (Approx-3):* *Approx-3* is a highly approximate design that computes all product bits in parallel by utilizing only the 6-input LUTs. Compared to *Approx-1* and *Approx-2* designs, it has a lower critical path delay and utilizes fewer LUTs for its implementation.
- *Approximate ternary adder:* To implement higher-order approximate multipliers from the 4 × 4 approximate multipliers, we present a novel approximate ternary adder. The adder eliminates the propagation of carries and utilizes only LUTs to compute the higher-order multiplier's final product.
- *Approximate signed multiplier (Booth-Approx):* We utilized the accurate *Booth-Opt* design presented in Chap. 3 and present an approximate radix-4 signed multiplier. Compared to the *Booth-Opt* design, *Booth-Approx* has reduced critical path delay and resource utilization.

The rest of the chapter is organized as follows. Section 4.2 presents a brief overview of some of the related state-of-the-art approximate multipliers. Section 4.3 presents three novel 4 × 4 approximate unsigned multipliers followed by their utilization in Sect. 4.4 to design higher-order multipliers. Section 4.5 presents the design of our proposed approximate signed multiplier. The implementation results and accuracy analysis of the presented multipliers are discussed in Sect. 4.6. Finally, Sect. 4.7 concludes the chapter.

4.2   Related Work

The design and exploration of approximate arithmetic circuits, particularly approximate multipliers, for error-resilient applications, have been an active area of research. An approximate multiplier can be implemented by exploring various approximations at the different stages of multiplications—partial products generations and their accumulation—and the multiplication algorithm itself. Here, we

provide a brief overview of some of the relevant state-of-the-art approximate multipliers. The authors of [3] have proposed an unsigned approximate 2×2 multiplier
for ASIC-based systems. Their presented multiplier produces a 3-bit output instead
of a 4-bit output by approximating the accurate product 9 (multiplying 3 with 3)
with 7. Their proposed technique results in reducing the overall complexity and
power consumption of the multiplier. Similarly, the authors of [2] have proposed
an unsigned approximate 2×2 multiplier by removing the logic for the least
significant product bit calculation. Their presented multiplier produces a maximum
error magnitude of 1 for three multiplication cases. Some works, such as [5–7], and
[8], have utilized approximations to implement fixed-width multipliers. A fixed-
width N $\times$ N multiplier produces a K-bit output, where $K < 2N$, by truncating the
computation of $2N - K$ least significant product bits. These works have proposed
various error compensation schemes to reduce truncation-induced errors in the final
product. Most works, such as [9] and [10], utilize approximate partial product
reduction trees to add the generated partial products for computing the approximate
product. The work presented in [9] utilizes a carry-maskable adder to decide
the length of the carry propagation while adding the partial product terms. The
authors of [10] have proposed FPGA-based approximate 3:2 and 4:2 compressors
for reducing the generated partial products to compute the product. The work
presented in [11] has utilized three different unsigned approximate 4×4 multiplier
designs to implement larger multipliers. The presented approximate 4×4 multipliers
utilize 6-input LUTs and the carry chains to implement various approximations
for computing the product bits. The authors of [12] have proposed approximate
radix-8 booth encoding to implement FPGA-based signed multipliers. As described
previously, the implementation of ASIC-based approximate operators on FPGAs
does not produce ASIC-like performance gains; therefore, towards this end, the
authors of [13] have utilized various machine learning models to explore ASIC-
based designs presented in [14] and identify Pareto-optimal designs for FPGA-based
systems.

4.3 Unsigned Approximate Multipliers

In this section, we first present three different designs of elementary 4×4
approximate multipliers. These designs are utilized in the following section to
implement higher-order, such as 8×8 and 16×16, approximate multipliers.

4.3.1 Approximate 4 × 4 Multiplier: Approx-1

Section 3.4, in Chap. 3, presents the design of an accurate unsigned multiplier,
referred to as *Acc*. The *Acc* multiplier combines the generation and mutual
addition of two consecutive partial product rows into one stage to produce a

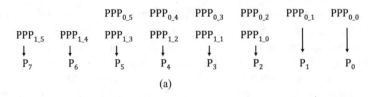

(a)

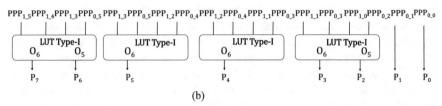

(b)

Fig. 4.1 Proposed approximate adder to implement approximate 4×4 multiplier *Approx-1*. (**a**) Grouping of PPPs. (**b**) LUT mapping of PPPs to compute approximate product

Processed Partial Product (PPP) row. To compute the final product, we had used an accurate ternary and binary adders-based partial product reduction tree for adding together the PPPs. However, we can also use an approximate reduction tree for adding accurate PPPs to compute the approximate product. We have used this technique to implement an approximate 4×4 multiplier, referred to as *Approx-1*, from accurate PPPs. The proposed approximate compressor eliminates the carry chains' utilization and adds all PPPs in parallel to compute the final product.

Figure 4.1a presents two rows of PPPs for a 4×4 multiplier, based on the structure described in Fig. 3.3. The least significant two PPP terms, PPP_{0_0} and PPP_{0_1}, do not require any further processing; therefore, these two bits are forwarded to the product bits P_0 and P_1, respectively. For the remaining PPP terms, we utilize approximate addition (a binary adder) as shown in Fig. 4.1b. Except for product bits P_6 and P_7, we have utilized LUT configuration Type-1 to compute the approximate product bits. The corresponding logic implemented by each configuration of LUT in the approximate adder is presented in Fig. 4.2. LUT configuration Type-I utilizes four PPP terms and produces two outputs. However, except for the least significant LUT Type-I, we utilize only the O6 output of the LUT to generate output. The least significant LUT configuration Type-I is utilized to compute accurate product bits P_2 and P_3. It receives four inputs (PPP_{0_2}, PPP_{1_0}, PPP_{0_3}, and PPP_{1_1}) and performs an *XOR* operation on the PPP_{0_2} and PPP_{1_0} inputs to compute accurate P_2. To compute P_3, the LUT first computes the carry out from the addition of PPP_{0_2} and PPP_{1_0} by performing an *AND* operation on these two bits. The computed carry-out signal is *XORed* with PPP_{0_3} and PPP_{1_1} to compute accurate P_3. For other instances of LUT Type-I, we utilize only the O6 signal of the LUT. These instances compute approximate product bits by estimating the carry out from the preceding locations and then *XORing* it with the other input bits. LUT configuration Type-II is used to compute the most significant product bits P_6 and P_7. For this purpose, the LUT estimates carry out from the preceding locations by performing

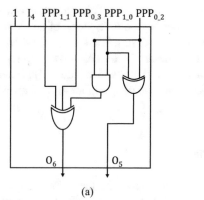 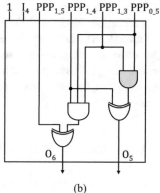

(a) (b)

Fig. 4.2 LUT logic for implementing proposed approximate adder for *Approx-1* multiplier. (**a**) LUT configuration Type-I. (**b**) LUT configuration Type-II

AND operation on PPP_{0_5} and PPP_{1_3}. The computed approximate carry is used to compute approximate product bits by performing *XOR* operations, as shown in Fig. 4.2.

As presented in Sect. 4.6 (Results and Discussion), the *Approx-1* multiplier offers a reduction in critical path delay and energy consumption of the multiplier when compared with the *Acc* multiplier. However, the detailed accuracy and performance characterization of *Approx-1* shows that utilization of approximate binary adders does not provide significant performance gains (LUT utilization, CPD, and PDP) compared with the Xilinx Vivado multiplier IPs [15]. In the following subsections, we apply various approximation techniques to present novel approximate multipliers that provide a better trade-off between output accuracy and the corresponding implementation performance gains.

4.3.2 Approximate 4 × 4 Multiplier: Approx-2

A performance-/area-optimized elementary multiplier module, targeted for FPGAs, should efficiently utilize the available 6-input LUT structure and the associated carry chains in a given FPGA. The 2 × 2 multipliers,[1] as used by [3] and [2], underutilize a 6-input LUT and therefore has been excluded from the list of potential design options for the elementary multipliers. The only two potential multiplier designs, which utilize all the inputs of a 6-input LUT, are 3 × 3 and 4 × 2 multipliers. However, a 3 × 3 multiplier is not a feasible option for the implementation of higher-order multipliers, e.g. 4 × 4 and 8 × 8 multipliers.[2] A 4 × 4 multiplier requires

---

[1] It should be noted that a row of PPPs in Fig. 4.1 is a also a 2 × 2 multiplier.

one 3×3, one 1×4, and one 3×1 multipliers [16]. This limited applicability of a 3×3 multiplier results in filtering it out from our selection of an elementary multiplier module. The only feasible elementary design is a 4×2 multiplier, which thoroughly utilizes lookup tables of state-of-the-art FPGAs. A 4×4 multiplier can be implemented using two instances of a 4×2 multiplier and an adder. *Approx-2* uses 4×2 multiplier as the elementary block for designing higher-order approximate multipliers.

4.3.2.1 Approximate 4 × 2 Multiplier

An accurate 4×2 multiplier generates a 6-bit output with the following optimized logic equations for $A(A_3A_2A_1A_0)$ and $B(B_1B_0)$ as multiplicand and multiplier, respectively:

$$P_0 = B_0A_0 \tag{4.1}$$

$$P_1 = B_1'B_0A_1 + B_1B_0'A_0 + B_1A_1'A_0 + B_0A_1A_0' \tag{4.2}$$

$$P_2 = B_1'B_0A_2 + B_1B_0'A_1 + B_0A_2A_1' + B_1A_2'A_1A_0' + B_1A_2A_1A_0 \tag{4.3}$$

$$P_3 = B_1'B_0A_3 + B_1B_0'A_2 + B_1A_3'A_2A_1' + B_0A_3A_2'A_1' \tag{4.4}$$

$$\qquad + B_1B_0A_3'A_2'A_1A_0 + B_0A_3A_2A_1 + B_0A_3A_1A_0'$$

$$P_4 = B_1B_0'A_3 + B_1A_3A_2'A_1' + B_1A_3A_2'A_0' + B_1B_0A_3'A_2A_1 \tag{4.5}$$

$$P_5 = B_1B_0A_3A_2 + B_1B_0A_3A_1A_0 \tag{4.6}$$

As P_0, P_1, and P_2 each depends on less than six shared variables, i.e., A_0, A_1, A_2, B_0, and B_1, any two of these three least significant product bits can be generated using a single 6-input LUT. The remaining four product bits will require four separate LUTs for implementation. An area- and energy-efficient approximation is to accommodate the six product bits in four LUTs, i.e., a single slice. Truncation of P_0 limits the output error to the least significant product bit and the final output accuracy to 75% with maximum error magnitude of ' 1 ' for all input combinations. Approximation of any other product bit results in a higher magnitude of error in the final output. The proposed approximate design of 4×2 multiplier uses 4 LUTs for its implementation by truncating ' P_0 ' and generating ' P_1 ' and ' P_2 ' by a single 6-input LUT.

---

[2] $2^x \times 2^x$, where $x \in \mathbb{N}^0$, multipliers are commonly used for applications in the domain of digital signal processing, computer vision, machine learning.

Fig. 4.3 4×4 multiplier
using 4×2 multipliers: P_{0X}
and P_{1X} represent result of
first and second 4×2
multipliers, respectively

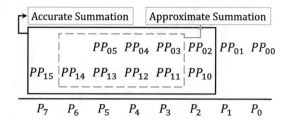

4.3.2.2 *Approx-2* Design

The approximate design of a 4×4 multiplier requires two 4×2 multipliers, consuming eight LUTs for generating partial products. For multiplicand $A(A_3 A_2 A_1 A_0)$ and multiplier $B(B_3 B_2 B_1 B_0)$, the first 4×2 multiplier takes $A(A_3 A_2 A_1 A_0)$ & $B(B_1 B_0)$ and the second 4×2 multiplier occupies $A(A_3 A_2 A_1 A_0)$ and $B(B_3 B_2)$ as input operands.

As shown by the black box in Fig. 4.3, the accurate summation of the approximate partial products generated by the two 4×2 multipliers requires the use of two carry chains.[3] Therefore, the approximate 4×4 multiplier, with accurate summation of partial products, requires 16 LUTs[4] (2 LUTs wasted by the second carry chain). Due to the truncation of PP_{00} and P_{10} in Fig. 4.3, this 4×4 multiplier implementation has an average relative error of 0.049 with an error probability of 0.375 for a uniform input distribution of all possible input combinations. However, the proposed design performs approximate addition along with FPGA-specific optimizations of second 4×2 multiplier and uses one single carry chain for partial products summation, as shown by the blue rectangle in Fig. 4.3. Our optimizations not only provides resources reduction but also significantly reduces the total number of error cases by having only *six* erroneous outputs. Our proposed optimization uses three LUTs for the implementation of required *carry-propagate* and *carry-generate* signals[5] to compute P_3, P_4, and P_5 product bits.

Since PP_{14} and PP_{15} share same six operands, our design does not compute PP_{14} and PP_{15} explicitly for subsequent addition by the carry chain. The proposed approach, as shown in Fig. 4.4, computes the respective *carry-propagate "Prop3"* and *carry-generate "Gen3"* signals for the computation of P_6 and P_7 directly from the multiplier and multiplicand bits by implicitly generating PP_{14} and PP_{15}. This implicit implementation of PP_{14} and PP_{15} saves one LUT as compared to their explicit computation. In order to improve the output accuracy, the recovered LUT is then assigned for the accurate realization of P_0 and P_2. Since the computation of P_3 is also dependent on the carry out from P_2, the corresponding LUT for P_3 besides using PP_{03} and PP_{11} also utilize A_0, B_2, and PP_{02} to resolve the effect of

---

[3] A carry chain is 4-bit wide in Xilinx 7 series FPGAs.

[4] Each position of a carry chain is controlled by a corresponding 6-input LUT.

[5] Computation of carry-propagate and carry-generate signals is defined in Sect. 2.2 in Chap. 2.

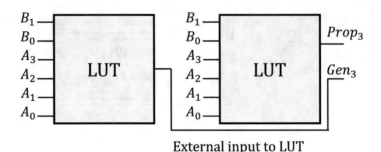

External input to LUT

Fig. 4.4 Implementation of *Gen₃* and *Prop₃* for P₆ and P₇ in *Approx-2* multiplier

Table 4.1 *Approx-2* multiplier error values

Multiplier	Multiplicand	Accurate product	Approximate product	Difference
5	15	75	67	8
6	7	42	34	8
6	15	90	82	8
7	15	105	97	8
13	13	169	161	8
15	5	75	67	8

the missing carry out from P_2. As carry-propagate and carry-generate signals cannot be ' 1 ' simultaneously, all the cases where A_0, B_2, PP_{02}, PP_{03}, and PP_{11} are all ' 1 ' concurrently will generate an error. To limit the error occurrences to a single product bit, P_3, we propose to correctly compute the carry-generate signal only. This decision limits the error to P_3 only with a fixed error magnitude of ' 8 '.

Tables 4.1 and 4.2 present the input operands with erroneous outputs and INIT values employed by each LUT along with input/output pins configuration, respectively. It is noteworthy that depending upon an application's input data, the proposed *Approx-2* multiplier may produce better result due to its asymmetric nature, and the values presented in Table 4.1 only show the maximum number of possible error occurrences for uniform distribution of all input cases. Our proposed multiplier does not generate erroneous outputs for highlighted inputs, in Table 4.1, with multiplier and multiplicand mutually swapped. For achieving better output quality results, the proposed approach suggests an initial analysis of input data, before multiplication, to decide operands for multiplier and multiplicand. The asymmetric nature of the proposed multiplier and the analysis of input data for achieving better output accuracy are further explored in Sect. 4.6.2.4.

Table 4.2 LUTs' input and output pins configuration for *Approx-2* multiplier

LUT	LUT input pins configuration						LUT output pins configuration	
	I5	I4	I3	I2	I1	I0	O6	O5
LUT_0	1	B_1	B_0	A_2	A_1	A_0	PP_{02}	$PP_{01} = P_1$
LUT_1	B_1	B_0	A_3	A_2	A_1	A_0	PP_{03}	
LUT_2	B_1	B_0	A_3	A_2	A_1	A_0	PP_{04}	
LUT_3	B_1	B_0	A_3	A_2	A_1	A_0	PP_{05}	
LUT_4	1	B_3	B_2	A_2	A_1	A_0	PP_{12}	PP_{11}
LUT_5	B_3	B_2	A_3	A_2	A_1	A_0	PP_{13}	
LUT_6	B_3	B_2	A_3	A_2	A_1	A_0	Gen_3	
LUT_7	1	1	PP_{02}	B_2	B_0	A_0	P_2	P_0
LUT_8	1	PP_{11}	PP_{03}	B_2	A_0	PP_{02}	$Prop_0$	Gen_0
LUT_9	1	1	1	1	PP_{12}	PP_{04}	$Prop_1$	Gen_1
LUT_{10}	1	1	1	1	PP_{13}	PP_{05}	$Prop_2$	Gen_2
LUT_{11}	B_3	B_2	A_3	A_2	A_1	A_0	$Prop_3$	

4.3.3 Approximate 4 × 4 Multiplier: Approx-3

Approx-3 multiplier design aims to reduce the total number of utilized LUTs, critical path delay, and power consumption of the implementation by reducing the number of computations' logic levels. The proposed design intends to thoroughly utilize the six inputs of a LUT and generates all product bits in parallel by eliminating carry chains' utilization. This design is also based on the basic multiplication algorithm shown in Fig. 3.1 in Chap. 3. For the parallel generation of all eight product bits, we group the corresponding partial product terms, as shown in Fig. 4.5a for multiplier $A \times B$. The grouping of partial product terms, represented by numbered boxes ⓪–⑤, is based on their respective locations. As shown by group ⓪ in Fig. 4.5a, the least significant two product bits, P_0 and P_1, depend on only four different bits (A_0, A_1, B_0, and B_1) of the 4-bit operands A and B. Therefore, product bits P_0 and P_1 can be computed accurately by a single 6-input LUT. Figure 4.5b shows the corresponding LUT configurations of group ⓪. The product bit P_0 is computed by performing logical AND operation on operand bit A_0 and B_0. Similarly, to compute product bit P_1, we perform logical AND operations on operand bits A_1 with B_0 and A_0 with B_1. The results of the two AND operations are summed together using a logical XOR operation to compute $P1$. As described in Sect. 2.2 in Chap. 2, the I_5 pin of the 6-input LUT must be set to "1" to enable the LUT to generate two separate outputs (P_0 and P_1 for LUT configuration ⓪). The I_4 pin of LUT ⓪ remains unutilized.

The computation of product bit $P2$, shown by group ① in Fig. 4.5a, depends on six different operands bits, i.e., A_0, A_1, A_2, B_0, B_1, and B_2. A single 6-input LUT can be used to perform the computation of group ①. As shown by the *unshaded* AND gates in LUT configuration ① in Fig. 4.5c, the respective bits are $ANDed$ together to compute the intermediate result. Further, the accurate computation of P_2

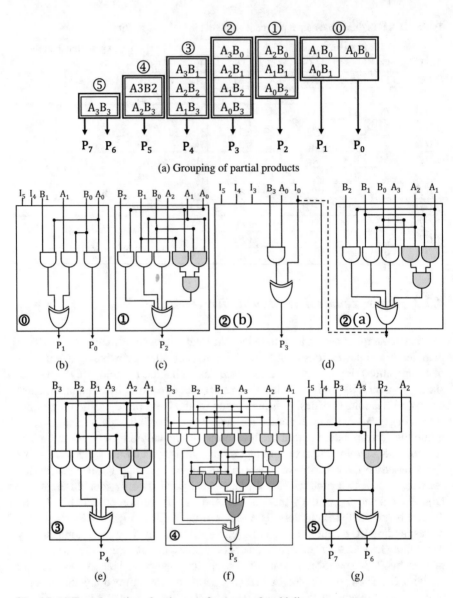

(a) Grouping of partial products

Fig. 4.5 LUT configuration of each group for *Approx-3* multiplier

also depends on the carry out from the preceding location (P_1). The deployed LUT configuration ① computes the required carry from inputs A_0, A_1, B_0, and B_1, as shown by the shaded gates in Fig. 4.5c. The outputs of all AND gates are summed together using an XOR operation to compute accurate product bit P_2.

Group ② is the only group encompassing *four* partial product terms with *eight* unique operands bits ($A_0 - A_3$ and $B_0 - B_3$). Therefore, to accommodate all operand bits and compute approximate product bit P_3, we have utilized two 6-input LUTs. As shown by LUT configurations in Fig. 4.5d, the LUT on the right, ②(a), receives *six* inputs to compute the sum of $A_3 B_0$, $A_2 B_1$, and $A_1 B_2$. Moreover, to improve computation accuracy, the LUT also estimates the carry out from the preceding group by utilizing A_1, A_2, B_0, and B_1, as shown by the shaded gates. A second LUT, ②(b), uses the output of the first LUT to compute the approximate product bit P_3. Three pins of the second LUT, i.e., I_3, I_4, and I_5, remain unused.

As group ③ also encompasses six operand bits, i.e., A_1, A_2, A_3, B_1, B_2, and B_3, the corresponding LUT configuration ③ has the same logic as LUT configuration ①. However, for computing the *carry out* from the preceding location, i.e., group ②, LUT configuration ③ has access to only A_1, A_2, B_1, and B_2. Therefore, LUT configuration ③ calculates the approximate *carry out*, shown by the shaded gates, and uses it to compute approximate product bit P_4.

Group ④, and the corresponding LUT configuration, is used to compute the approximate product bit P_5. This group encompasses four operand bit, i.e., A_2, A_3, B_2, and B_3, as shown in Fig. 4.5a. A single 6-input LUT can accommodate these four bits to compute an approximate product bit P_5. However, to improve the output accuracy of the P_5, we utilize the two remaining pins of the 6-input LUT by providing operand bits A_1 and B_1 to predict the carry out from the preceding group (i.e., group ③). It should be noted that group ② also encompasses two minterms, $A_2 B_1$ and $A_1 B_2$, that can also generate a carry. The proposed configuration for LUT ④ explores the various possibilities for an input carry from the preceding group location to compute approximate P_5. These possibilities are described in Eqs. 4.7 and 4.8 and implemented using the shaded gates in Fig. 4.5f. Finally, the computed approximate carry is added with the minterms $A_2 B_3$ and $A_3 B_2$ to calculate approximate product bit P_5:

$$c_0 = A_2 B_1 \cdot A_1 B_2 \qquad (4.7)$$

$$approx.\ carry\_out = A_3 B_1 \cdot A_2 B_2 + A_3 B_1 \cdot A_1 B_3 + A_2 B_2 \cdot A_1 B_3 \qquad (4.8)$$
$$+ c_0 \cdot A_3 B_1 + c_0 \cdot A_2 B_2 + c_0 \cdot A_1 B_3$$

Group ⑤ is used to compute the most significant product bits P_6 and P_7. Product bit P_7 is the output carry obtained after adding $A_3 B_3$ with the carry out from the preceding location. Therefore, LUT configuration ⑤ also reads A_2 and B_2 to compute the approximate carry out from group ④, as shown by the 4-input shaded AND gate. This approximate carry is then utilized in the computation of product bits P_6 and P_7. This group also generates two different outputs; therefore, the I_5 pin of LUT ⑤ must be provided with a constant 1. The I_4 pin of the LUT remains unconnected.

4.4   Designing Higher-Order Approximate Unsigned Multipliers

We have used the modular approach for implementing unsigned approximate multipliers using sub-multipliers. As described in Fig. 1.6, multiple sub-multipliers are added together to implement a larger multiplier in this technique. For example, Sect. 4.3.2 presented an approximate 4×4 multiplier design using two instances of 4×2 multipliers. In general, any higher-order multiplier can be implemented using multiple smaller multipliers. In this book, we refer to the output generated by each sub-multiplier as a Sub-product (SP). Various accurate and approximate adders can be utilized to add together the SPs of sub-multipliers to obtain the final accurate/approximate product. In the following sections, we show various designs of approximate 8×8 multipliers implemented using proposed approximate 4×4 multipliers and accurate and approximate adders.

4.4.1   Accurate Adders for Implementing 8 × 8 Approximate Multipliers from 4 × 4 Approximate Multipliers

Utilizing the modular approach, we have used four instances of our proposed approximate 4×4 multipliers and accurate addition to implement approximate 8×8 multipliers. We can also use accurate binary addition (addition having two operands) to add the four SPs using three binary adders. The first two adders will add the four SPs, and the third adder will be used for adding together the result of the other two adders. However, an analysis of the SPs arrangement reveals that only three out of four SPs can be added together at any position. Therefore, instead of binary adders, we utilize ternary adders[6] (addition having three operands) for adding the SPs. Utilizing the proposed approximate multipliers *Approx-1*, *Approx-2*, and *Approx-3*, we implement higher-order multipliers, referred to as *Approx-1*-a, *Approx-2*-a, and *Approx-3*-a, respectively, in this book. For example, to implement an 8×8 *Approx-1*-a multiplier, the $SB<4>-SB<7>$ from $A_L \times B_L$, $SB<0>-SB<3>$ from $A_H \times B_L$ and $SB<0>-SB<3>$ from $A_L \times B_H$ are added in one single step to produce final product bits P_4-P_7 for an 8×8 multiplier. The $O5$ output of the fourth LUT and the Cout of the carry chain in Fig. 3.7 are routed to the next slice for generation of higher-order product bits. The same process can be repeated for the implementation of arbitrary sizes of higher-order multipliers.

The modular approach provides a broader design space by using various accurate/approximate sub-multipliers to implement a higher-order accurate/approximate multiplier. For example, we have used the unsigned accurate 4×4 multiplier *Acc* presented in Sect. 3.4 and the unsigned approximate 4×4 designs *Approx-1*,

---

[6] Ternary adders are introduced in Sect. 3.4 and presented in Fig. 3.7.

Approx-2, and *Approx-3* to design 256 8 × 8 approximate multipliers with different accuracy and performance metrics. For this purpose, we have used the accurate ternary adders to add the SPs of the sub-multipliers. The utilization of approximate adders to add the generated SPs can further extend the total design space with new design points.

4.4.2 Approximate Adders for Implementing Higher-order Approximate Multipliers

We also present a novel approximate adder for the approximate addition of the results (SPs) of accurate/approximate sub-multipliers to implement higher-order approximate multipliers. Our proposed 6-input LUTs-based adder adds the results of three sub-multipliers simultaneously without using the associated carry chain in a slice. The proposed approximate adder further improves the performance of the realized higher-order multipliers by reducing the critical path of the addition process.

In this design, we do not estimate the carry out from the preceding locations, and a product bit is computed by performing XOR operation on the respective SP bits. For example, to implement an 8 × 8 approximate multiplier using 4 × 4 sub-multipliers, Fig. 4.6 presents the approximate ternary adder logic implemented by each 6-input LUT to compute the approximate product bits. Further to reduce the overall resource utilization of the approximate addition, we do not add the MSBs of the most significant SP. For example, Fig. 4.7 shows the implementation of an approximate 8 × 8 multiplier using accurate/approximate 4 × 4 sub-multipliers. Each blue box represents an instance of the proposed approximate adder implemented using a 6-input LUT. The same technique is utilized for implementing other higher-order multipliers. To compare the impact of the proposed approximate adder on the performance and accuracy of higher-order multipliers, we have used

Fig. 4.6 Proposed approximate adder for implementation of higher-order multipliers: LUT-based representation

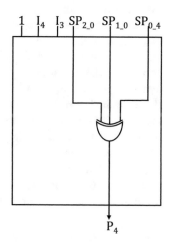

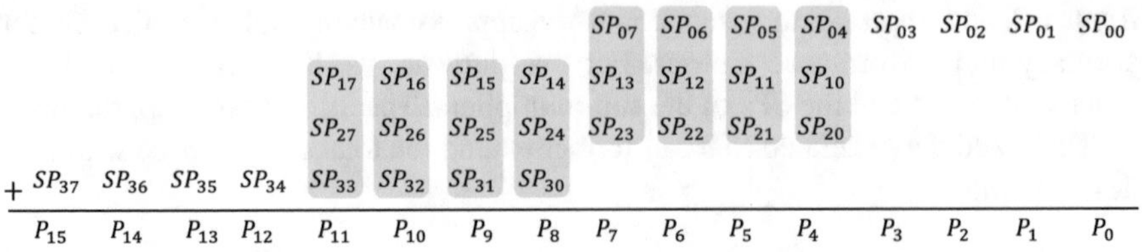

					P10	P9	P8	P7	P6	P5	P4	P3	P2	P1	P0
								SP_{07}	SP_{06}	SP_{05}	SP_{04}	SP_{03}	SP_{02}	SP_{01}	SP_{00}
							SP_{17}	SP_{16}	SP_{15}	SP_{14}	SP_{13}	SP_{12}	SP_{11}	SP_{10}	
						SP_{27}	SP_{26}	SP_{25}	SP_{24}	SP_{23}	SP_{22}	SP_{21}	SP_{20}		
+	SP_{37}	SP_{36}	SP_{35}	SP_{34}	SP_{33}	SP_{32}	SP_{31}	SP_{30}							

P_{15}	P_{14}	P_{13}	P_{12}	P_{11}	P_{10}	P_9	P_8	P_7	P_6	P_5	P_4	P_3	P_2	P_1	P_0

Fig. 4.7 8×8 approximate multiplier and its approximate summation

the approximate 4×4 *Approx-1*, *Approx-2*, and *Approx-3* multiplier designs to implement higher-order approximate multipliers, referred to as *Approx-1*-p, *Approx-2*-p, and *Approx-3*-p, respectively, in this book. Further, utilizing *Acc*, *Approx-1*, *Approx-2*, *Approx-3*, and the proposed approximate ternary adder, we generate another set of 256 approximate 8×8 multipliers. The accuracy and performance results of these multipliers are compared with the 256 approximate multipliers implemented using accurate addition for SPs of *Acc*, *Approx-1*, *Approx-2*, and *Approx-3* sub-multipliers.

4.5 Approximate Signed Multipliers: *Booth-Approx*

Section 3.7 presented a generic area-optimized signed multiplier architecture— *Booth-Opt*. Utilizing *Booth-Opt* as a base architecture, we perform a detailed analysis of the possible trade-offs between final output accuracy, latency, and energy gains for different sizes of multipliers. For each multiplier, we have used multiple uniform distributions of all input combinations to estimate power dissipation in each of the instantiate LUTs of the respective implementations. Furthermore, to compute the dynamic power of each multiplier, the testbench of each multiplier utilizes a clock period that is the two times critical path delay of the respective design under test. We also analyze and identify the LUTs and carry chains contributing to the overall critical path delay of each multiplier.

Figure 4.8 shows the top five power-consuming elements and the top five critical paths highlighted in the schematic of a 6×6 *Booth-Opt* multiplier for multiple uniform distributions of all input combinations. Our analysis reveals that the first two LUTs (LSBs) in each partial product row contribute more to the multiplier dynamic power consumption and critical path delay. For instance, the first LUT (LSB) in the last partial product row of the 6×6 *Booth-Opt* multiplier contributes the most ($50.943\,\mu W$) to the total dynamic power consumption ($865\,\mu W$). Similarly, for an 8×8 multiplier, the highest power dissipation of $170\,\mu W$ is observed for the first LUT compared to $84\,\mu W$ for the fourth LUT in the fourth (most significant) partial product row. For the 6×6 multiplier, the first LUT (LSB) in each partial product row (total three rows) contributes to the top five worst critical path delays. Similarly, for an 8×8 multiplier, three first-placed LUTs in the partial product

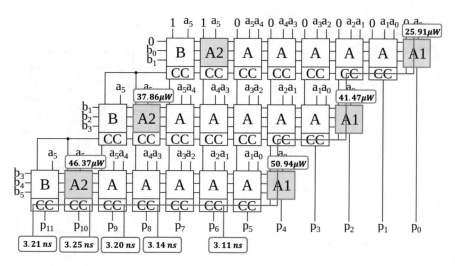

Fig. 4.8 Schematic of 6 × 6 *Booth-Opt* multiplier showing top five critical paths per output (in blue) and top five most power-consuming elements (in red)

rows (total four rows) contribute to all of the top five worst critical path delays. Therefore, approximating the functionalities of the first two LUTs in each partial product row can lead to significant power and latency gains. It must be noted that the power dissipation is due to the switching in the LUTs and the routing power of the carry signal generated from such LUTs and used by the other LUTs. Since the LSBs switch more often and the resulting carry signal is propagated across most other blocks, such blocks dissipate the most dynamic power. We recommend the following modifications/suggestions to an N×N base architecture (*Booth-Opt*) for achieving a latency and power-optimized approximate signed multiplier.

- We propose truncation of the first LUT, for the LSB, in each partial product row of an N×N multiplier to static "0." This truncation results in a significant decrease in dynamic power consumption.
- To approximate the output of second LUT in each partial product row, we propose LUT configuration Type-Am, shown in Fig. 4.9. LUT Type-Am does not use the associated carry chain and predicts the missing input carry, using signal "i," to generate an approximate output. The detailed error analysis of the outputs generated by second LUT in each partial product row of our base architecture reveals that in the absence of an input carry, most errors are generated for the Booth's encoding $\overline{1}, \overline{2}, 2$. To reduce the number of these wrong outputs, LUT Type-Am predicts the missing input carry as constant "1" for Booth's encoding $\overline{1}, \overline{2}, 2$ and uses it for computing the approximate output.
- As shown by Verma et al. [17], the chances of errors in higher-order output bits, produced by an initial incorrect input carry, decrease with the increasing length of the carry chain. Our proposed approach recommends a constant "0/1" input

Fig. 4.9 LUT configuration
Am for implementing
Booth-Approx multiplier

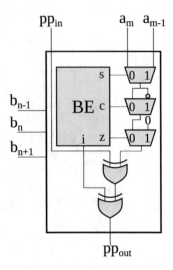

carry to the first LUT of Type-A in a partial product row. The exhaustive error analysis of smaller multipliers, such as an 8×8 multiplier, reveals that providing a constant "1" as input carry results in a decreased relative error in the final output. However, as the most significant partial product row has the maximum contribution in the accuracy of the final product, the carry generation in the most significant partial product row should remain unaffected. Therefore, the most significant partial product row utilizes a LUT Type-A1 (denoted by A1*) only for carry generation and the pp_{out} signal of A1* is truncated to constant "0."

Utilizing these guidelines, Fig. 4.10 shows an example of *Booth-Approx* by describing the architecture of a 6×6 approximate signed multiplier. Our proposed approximate signed multiplier resource utilization can be estimated by Eq. 4.9:

$$LUTs \ for \ M \times N \ multiplier \ = (M + 1) \times \left\lceil \frac{N}{2} \right\rceil + 1 \qquad (4.9)$$

To evaluate the impact of the proposed optimizations on the performance metrics of the multiplier, we perform power and critical path delay analysis of the 6×6 *Booth-Approx* multiplier. For this purpose, the testbench provides a uniform distribution of all input combinations to the multiplier using a clock period double of the critical path delay of the approximate multiplier. Figure 4.11 shows the top five dynamic power-consuming elements and the top five outputs having the worst critical path delay. Compared to the top five power-consuming elements of *Booth-Opt* (shown in Fig. 4.8), the overall power consumption of the *Booth-Approx* is significantly reduced. The LUTs at the least significant locations are no longer the major contributors to the total power consumption. Further, compared to *Booth-Opt*, the power distribution of the *Booth-Approx* also includes a carry chain element among the most power-consuming elements. Compared to the 6-input LUTs, the carry chain elements provide a high-performance and energy-

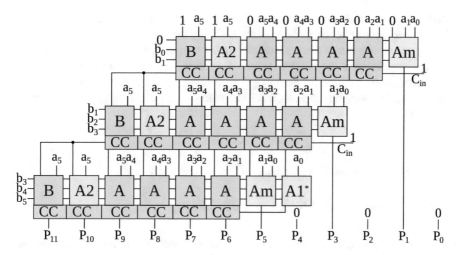

Fig. 4.10 A 6×6 approximate signed multiplier implementation (*Booth-Approx*)

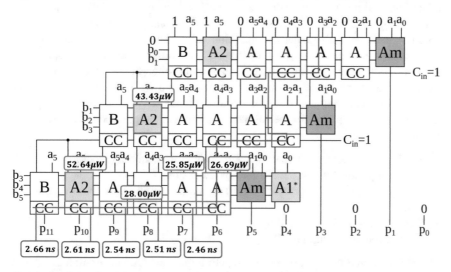

Fig. 4.11 Schematic of 6 × 6 *Booth-Approx* multiplier showing top five critical paths per output (in blue) and top five most power-consuming elements (in red)

efficient fixed implementation of carry-lookahead adders. The power analysis of the 6 × 6 *Booth-Approx* shows that most of the LUTs consume less power than the energy-efficient carry chains. The proposed optimizations/approximations have also significantly reduced the overall critical path delay of the *Booth-Approx* multiplier. It should be noted that the proposed LUT configuration Type-A provides the carry-generate signal for the associated carry chain as the external bypass signal. As shown by the critical path delay analysis in Fig. 4.11, the external bypass signal, in the second partial product row, lies on the critical path of all outputs. Compared to the accurate multiplier analysis in Fig. 4.8, the analysis presented in Fig. 4.11 is performed on a higher clock frequency.[7] Utilizing the same clock frequency for both implementations shows further reduction in the overall dynamic power consumption of the approximate circuit. For example, utilizing two times critical path delay of the accurate multiplier (*Booth-Opt*) as a clock period for the testbench of the approximate multiplier shows a $7.57\mu W$ reduction in the most power-consuming element.

4.6   Results and Discussion

4.6.1   Experimental Setup and Tool Flow

To evaluate the performance metrics of various proposed approximate multipliers, we have used the experimental setup presented in Sect. 3.9.1. This setup implements all designs multiple times with different critical path delay constraints for the Virtex-7 family FPGA. Further, it uses Vivado Simulator and Power Analyzer tools to compute each design's power consumption value. The accuracy of the proposed approximate multipliers has been computed for multiple uniform distributions of all input combinations. Moreover, the C++ and Python-based behavioral models of proposed approximate multipliers are also deployed in various applications from image processing and machine learning domains. We compare the proposed approximate multipliers for performance gains and output accuracies with the following designs:

- S1 [2]: utilizes an approximate 2×2 multiplier as a building block.
- S2 [3]: approximate 2×2 multiplier-based design.
- EvoApprox [14, 18]: library of 8-bit approximate multipliers (both signed and unsigned).
- Precision-reduced 8×8 multipliers
- Xilinx accurate multiplier IP[15]

We have explored three different methods to implement a precision-reduced 8×8 multiplier.

1. Truncating multiple LSBs in the final product of an accurate M $\times$ N multiplier to "0" after performing multiplication. However, this method has no impact on the synthesis tool in reducing the overall resource utilization, critical path delay, and energy consumption. The generated precision-reduced multiplier offers the performance metrics of an accurate M $\times$ N multiplier.
2. Removing the logic for the computation of some LSBs in the final product of an M $\times$ N multiplier and generating the rest of the product bits accurately. However,

---

[7] The achieved critical path delay determines the clock frequency of the testbench in our setup.

the accurate computation of product MSBs in the absence of propagated information (i.e., output carries) from the removed product LSBs significantly increases the multiplier's overall complexity and resource utilization. In such an implementation, each product bit is computed independently. This design is denoted by *Mult(Number of product bits, Number of truncated bits)* in this book. For instance, implementing a precision-reduced $Mult(16, 4)$ multiplier (i.e., 12 accurate MSBs and 4 removed LSBs) requires 350 LUTs—an increase of 301% over the 88 LUTs of Vivado's area optimized 8×8 design.

3. Truncating the LSBs of the input operands before multiplication and then utilizing a lower bit-width accurate multiplier for multiplication. Depending upon the number of truncated bits in the operands, the computed product is shifted to calculate the final product. This design is denoted by *P(Operand bit-width, Number of truncated bits)*. For example, the $P(8, 1)$ and $P(8, 2)$ precision-reduced multipliers truncate one and two bits of each operand, respectively. These precision-reduced operands are then utilized by corresponding smaller accurate multipliers (7×7 and 6×6) to compute the product, which is shifted by an appropriate number of bits (2-bits and 4-bits) to calculate the final product.

In the following sections, we evaluate the efficacy of our proposed unsigned and signed multipliers by comparing their accuracy and implementation performance metrics with state-of-the-art accurate and approximate multipliers.

4.6.2 Evaluation of the Proposed Approximate Unsigned Multipliers

4.6.2.1 Performance Characterization of Designed Multipliers

Table 4.3 presents the implementation results of our proposed approximate unsigned multipliers. The N×N *Acc*-p multipliers use four instances of $\frac{N}{2} \times \frac{N}{2}$ accurate multiplier Acc[8] to generate sub-products and the proposed approximate ternary adder to add them. As discussed previously, the *Approx-1*-a, *Approx-2*-a, and *Approx-3*-a multipliers are implemented by using four instances of 4x4 *Approx-1*, *Approx-2*, and *Approx-3*, respectively, and adding the resulting sub-products by deploying the accurate ternary adder. Similarly, *Approx-1*-p, *Approx-2*-p, and *Approx-3*-p multipliers utilize approximate ternary adder for adding the sub-products to compute the final product. The proposed 4×4 approximate multipliers provide reduced Critical Path Delay (CPD) and energy consumption (PDP) than the *Acc* multiplier. *Approx-3* multiplier provides the minimum CPD and energy consumption among all presented multipliers by eliminating the carry chain utilization and parallel generation of all product bits. Further, *Approx-3* utilizes only

---

[8] Accurate multiplier architecture presented in Sect. 3.4.

Table 4.3 LUTs, CPD, and PDP results of proposed unsigned multipliers

Multiplier size	Design	Area [LUTs]	CPD [ns]	PDP [pJ]
4 × 4	Acc	12	2.016	1.127
	Approx-1	12	1.704	1.100
	Approx-2	12	1.564	0.649
	Approx-3	7	0.805	0.284
	Approx-1-a	57	3.479	6.506
	Approx-2-a	57	3.130	4.732
	Approx-3-a	37	2.527	2.754
8 × 8	Acc-p	56	2.388	5.886
	Approx-1-p	56	2.194	5.148
	Approx-2-p	56	1.982	3.546
	Approx-3-p	36	1.409	1.824
	Approx-1-a	245	5.159	31.558
	Approx-2-a	245	4.979	26.495
	Approx-3-a	165	4.075	16.775
16 × 16	Acc-p	224	4.301	28.448
	Approx-1-p	240	3.204	22.940
	Approx-2-p	240	2.375	16.155
	Approx-3-p	160	1.968	8.687

seven LUTs to implement a 4 × 4 multipliers. The better performance (LUT utilization, CPD, and PDP) of *Approx-3*-based multipliers is also valid for higher-order multipliers. For example, compared to the 8 × 8 *Acc* multiplier, *Approx-3*-a offers 28.8%, 32.7%, and 54.8% reductions in the total utilized LUTs, CPD, and PDP, respectively. However, as described in the following section, the elimination of the carry chain and approximate computing of most of the product bits for the *Approx-3* multiplier results in reduced output accuracy of the multiplier.

The proposed *Approx-2* multiplier has slightly higher resource utilization, CPD, and PDP than the *Approx-3* multiplier; however, it offers better output accuracy than *Approx-3*. Compared to the 4 × 4 *Acc* multiplier, *Approx-2* offers a 22.4% and 42.4% reduction in the CPD and PDP, respectively. Similarly, compared to the 8 × 8 *Acc* multiplier, 8 × 8 *Approx-2*-a reduces the CPD and PDP by 16.6% and 22.4%, respectively.

As shown by the results, the approximate summation of partial products has helped in significantly reducing the latency and energy consumption of all multipliers. For example, compared to the 8 × 8 *Acc* multiplier presented in Chap. 3, the *Acc*-p, *Approx-1*-p, *Approx-2*-p, and *Approx-3*-p show 36.4%, 41.5%, 47.2%, and 62.4% reduction in the CPD, respectively. Similarly, the *Acc*-p, *Approx-1*-p, *Approx-2*-p, and *Approx-3*-p multipliers show a reduction of 3.4%, 15.59% 41.8%, and 70.0% in energy consumption (PDP), respectively, when compared with the *Acc* multiplier.

Table 4.4   Error analysis of *Approx-1*-based 8×8 approximate multipliers

Error	Approximate architectures						
description	*Approx-1*-a	*Approx-1*-p	S1 [2]	S2 [3]	Mult(16,4)	P(8,1)	P(8,2)
Maximum error magnitude	23120	27,488	7225	14,450	15	509	1521
Average error	1734	3103.70	1354.69	903.12	6.50	127.25	380.25
Average relative error	0.088	0.19	0.14	0.032	0.0037	0.026	0.069
Error occurrences	29222	54800	53,375	30,625	53,248	48,896	61,056
Maximum error occurrences	1	1	31	1	2048	1	1

Table 4.5   Error analysis of *Approx-2*-based 8×8 approximate multipliers

Error	Approximate architectures							
description	Ca	Cc	Acc_app	S1 [2]	S2 [3]	Mult(16,4)	P(8,1)	P(8,2)
Maximum error magnitude	2312	8288	8160	7225	14,450	15	509	1521
Average error	54.19	1592.26	1579.12	1354.69	903.12	6.50	127.25	380.25
Average relative error	0.0029	0.13	0.13	0.14	0.032	0.0037	0.026	0.069
Error occurrences	5482	52,731	52,437	53,375	30,625	53,248	48,896	61,056
Maximum error occurrences	14	1	2	31	1	2048	1	1

4.6.2.2   Error Analysis of Proposed Approximate Multipliers

Tables 4.4, 4.5, and 4.6 present an error analysis of proposed approximate multipliers in comparison with the state-of-the-art approximate multipliers and precision-reduced 8×8 multipliers Mult(16,4), P(8,1), and P(8,2). As shown in Tables 4.4 and 4.6, the elimination of the carry propagation has reduced the output accuracy of *Approx-1* and *Approx-3* multipliers. For example, compared to the state-of-the-art approximate multipliers, the *Approx-1*- and *Approx-3*-based 8×8 multipliers have higher maximum error and average error values. However, *Approx-1*-a multiplier is better than state-of-the-art multipliers in terms of the total number of error occurrences (error probability). Similarly, *Approx-3*-a also has a smaller error probability than most of the other multipliers. Further, for all *Approx-1*- and *Approx-3*-based multipliers, the total occurrences of maximum error values are also only one. It should be noted that despite having high values for various statistical error metrics, an approximate multiplier can still produce acceptable quality application-

Table 4.6 Error analysis of *Approx-3*-based 8×8 approximate multipliers

Error description	Approximate architectures						
	Approx-3-a	*Approx-3*-p	S1 [2]	S2 [3]	Mult(16,4)	P(8,1)	P(8,2)
Maximum error magnitude	27,744	31,736	7225	14,450	15	509	1521
Average error	3251.25	4440.75	1354.69	903.12	6.50	127.25	380.25
Average relative error	0.16	0.25	0.14	0.032	0.0037	0.026	0.069
Error occurrences	45,751	56,929	53,375	30,625	53,248	48,896	61,056
Maximum error occurrences	1	1	31	1	2048	1	1

level results for error-resilient applications. In the following Sect. 4.6.2.4, we have used the proposed approximate multipliers in two image processing applications to evaluate their impact on the output quality.

The proposed multiplier *Approx-2*-a outperforms all approximate multipliers in terms of maximum error magnitude, error occurrences, and maximum error occurrences, as shown in Table 4.5. The approximate multiplier *Approx-2*-p has a higher maximum error magnitude compared to the state-of-the-art S1[2]; however, the maximum error occurs only once for *Approx-2*-p, while it occurs 31 times for S1[2]. Table 4.5 also shows the error analysis of the 8 × 8 *Acc*-p multiplier. Compared to the *S1* [2], *Acc*-p has a higher maximum error magnitude and average error value; however, *Acc*-p offers a lower average relative error, total error occurrences, and maximum error value occurrences than *S1*. Compared to *S2* [3], *Acc*-p has a lower maximum error value.

The precision-reduced Mult(16,4) has highest number of maximum error occurrences. Regardless of its low average relative error, its high resource utilization, 350 LUTs, filters it out in Pareto analysis. The precision-reduced P(8,1) and P(8,2) multipliers offer reduced utilization of resources due to the use of smaller accurate multipliers. However, P(8,1) and P(8,2) have the highest number of error occurrences. The average error and average relative error of P(8,1) and P(8,2) are also higher than the *Approx-2*-a multiplier.

To further analyze the high output accuracy of *Approx-2*-based multipliers, we explore the erroneous bit values with their effect on the final output and the frequency of error occurrences. A similar analysis can also be done for *Approx-1*- and *Approx-3*-based multipliers. Figure 4.12 represents the normalized bit accuracy histograms and the number of unique error occurrences for *Approx-2*-based multipliers. As shown in Fig. 4.12a, the 4 × 4 *Approx-2* multiplier has errors only in product bit 3 (P_3) for all 256 input combinations. For a total of 6 multiplications, the *Approx-2* wrongly computes P_3 to be 0. These input combinations are described in Table 4.1. The accuracy bit histogram of *Approx-2*-a, in Fig. 4.12b, also shows that

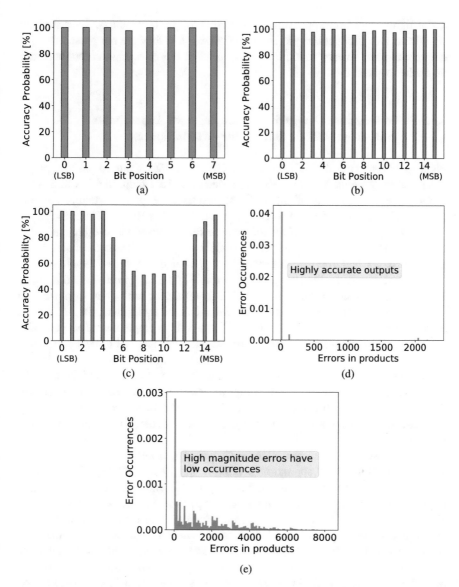

Fig. 4.12 Probability of error in individual product bits of *Approx-2*-based multipliers: (**a**, **b**), and (**c**) show the normalized bit histograms of different multipliers. (**d**) and (**e**) show the histogram of unique errors in different multipliers. (**a**) 4×4 *Approx-2*. (**b**) 8×8 *Approx-2*-a. (**c**) 8×8 *Approx-2*-p. (**d**) 8×8 *Approx-2*-a. (**e**) 8×8 *Approx-2*-p

the product bits are computed correctly for most multiplications. Further, as shown in Fig. 4.12d, the *Approx-2*-a has only a few distinct errors. Figure 4.12c and e show the accuracy bit histogram and number of distinct errors for the 8×8 *Approx-2*-p multiplier. The low probability of getting accurate bit values for *Approx-2*-p is due to the highly inaccurate approximate addition of the partial products. Since the least significant four product bits are already finalized, and they are not passed through the approximate ternary adder, these four product bits have high output accuracy than other product bits.

4.6.2.3 Performance Comparison of the Proposed Approximate Multipliers with the State-of-the-Art Multipliers

Figure 4.13 compares the LUT utilization, CPD, and PDP of our proposed approximate multipliers with Vivado area- and speed-optimized multiplier IPs and state-of-the-art approximate multipliers. These results have been normalized to the corresponding results of Vivado area-optimized multiplier IP [15]. The actual value for every implementation performance metric for the Vivado area-optimized IP is also shown above the normalized performance bar of the IP. Please note that *Approx-1*-a, *Approx-2*-a, and *Approx-3*-a designs are represented as *Approx-1*, *Approx-2*, and *Approx-3*, respectively, for clarity. As shown in Fig. 4.13a, *Approx-1*- and *Approx-2*-based designs tend to utilize more LUTs than the *Acc* multiplier. Compared to the modular designs of *Approx-1*- and *Approx-2*-based multipliers, the *Acc* design is based on array-based implementation. The modular implementation generally occupies more resources; however, these designs offer reduced critical path delay and corresponding energy consumption. For example, as shown in Fig. 4.13b and c, the *Approx-1*- and *Approx-2*-based designs offer reduced CPD and PDP than the corresponding *Acc* multipliers. The proposed *Approx-3*-based designs outperform all designs in LUT utilization, CPD, and PDP. The better performance of the *Approx-3*-based designs is due to the elimination of the carry chain logic and generating all product bits in parallel.

The proposed approximate designs always occupy fewer LUTs when compared with the state-of-the-art approximate designs S1 [2] and S2 [3]. Compared to the Vivado area- and speed-optimized IPs, *Approx-1*- and *Approx-2*-based designs require fewer LUTs for smaller multipliers, e.g., 4 × 4 and 8 × 8. However, for 16 × 16 multipliers, these designs require slightly more resources than the IPs. Nevertheless, the *Approx-3*-based designs always require fewer LUTs than the Vivado IPs. For example, compare to the Vivado area-optimized 8 × 8 multiplier IP, *Approx-1*-a, *Approx-2*-a, *Approx-3*-a, *Approx-1*-p, *Approx-2*-p, and *Approx-3*-p offer 3.3%, 3.3%, 37.2%, 5.0%, 5.0%, and 38.9% reduction in the total utilized LUTs, respectively.

Compared to the S2 design, the *Approx-1*-a, *Approx-2*-a, and *Approx-3*-a designs have inconsiderably higher CPD. The insignificant increase in the CPD of the multiplier is mainly due to the increased CPD of the utilized accurate ternary adders. Therefore, the approximate ternary addition-based designs (*Approx-1*-p, *Approx-2*-p, and *Approx-3*-p) have reduced CPD when compared with S2. For example, compared to the 8 × 8 S2 multiplier, *Approx-1*-p, *Approx-2*-p, and *Approx-3*-p reduce the CPD by 5.6%, 14.7%, and 39.4%, respectively. Similar results are observed when the proposed multipliers are compared with Vivado IPs. For example, *Approx-*

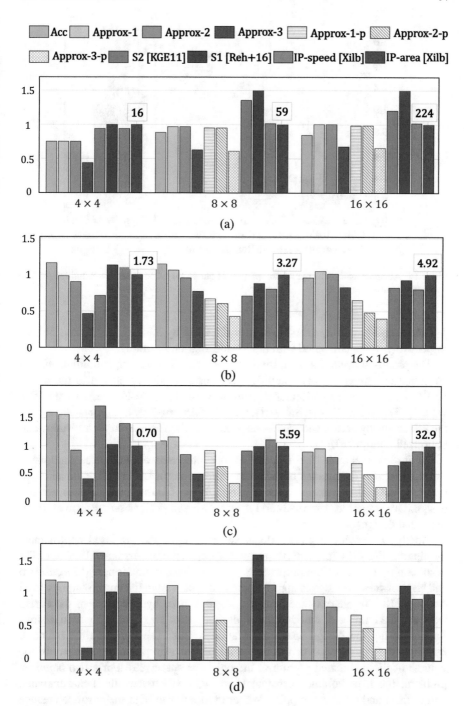

Fig. 4.13 LUT utilization, CPD, and energy consumption of accurate and approximate unsigned multipliers. Results are normalized to the corresponding results of Vivado area-optimized multiplier IP [15]. (**a**) LUTs. (**b**) CPD. (**c**) PDP. (**d**) PDP × LUTs

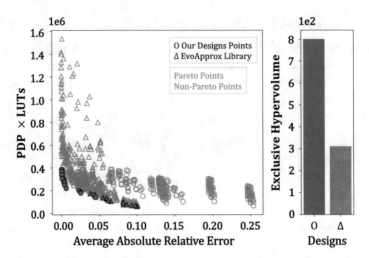

Fig. 4.14 Accuracy and performance comparison of our proposed unsigned 8 × 8 multipliers with EvoApprox multipliers [18]

1-p, *Approx-2*-p, and *Approx-3*-p reduce the CPD by 33.0%, 39.4%, and 56.9%, respectively, when compared with the Vivado area-optimized IP.

The proposed *Approx-2*- and *Approx-3*-based designs have the minimum energy consumption among all presented designs. For example, compared to the 8 × 8 Vivado area-optimized multiplier IP, *Approx-3*-a and *Approx-3*-p decrease the PDP value by 67.3% and 50.7%. Similarly, the 8 × 8 *Approx-2*-a and *Approx-2*-p reduce the PDP value by 36.6% and 15.4%, respectively, when compared with Vivado area-optimized multiplier IP.

To further elaborate the efficacy of our proposed approximate multipliers, Fig. 4.13d shows the product of the normalized values of LUTs utilization and PDP (PDP × LUTs). A smaller value of the PDP × LUTs metric denotes a better design. Our proposed *Approx-2*- and *Approx-3*-based designs outperform all other presented designs.

Finally, to provide a more exhaustive analysis of the proposed approximate multipliers, Fig. 4.14 compares the average absolute relative error and PDP × LUTs of all configurations of the proposed 8 × 8 approximate multipliers (512 designs) and state-of-the-art unsigned approximate multipliers from EvoApprox library (504 designs) [18]. To quantify the analysis, we have used hypervolume of the Pareto front and the number of non-dominated design points, two commonly used metrics for comparing the design space exploration results. The hypervolume indicator measures the significance of the non-dominated design points by computing the volume of the dominated portion of the objective space [19]. For a two-objective problem, the hypervolume corresponds to the area between the non-dominated Pareto front and a reference point. We aim to minimize both the absolute average relative error and PDP × LUTs for our current work; therefore, the reference point

Table 4.7 Average PSNR of ten images from [20] for image blending application

Multipliers	Approx-1-a	Approx-1-p	Approx-2-a	Approx-2-p	Approx3-a	Approx3-p
Average PSNR	38.6	30.4	51.9	31.6	32.5	29.4

comprises of the maximum of both the metrics across all Pareto front points under consideration.

For the 504 8 × 8 multiplier implementations provided by the EvoApprox library, 16 designs lie on the Pareto surface. However, our proposed approximate multipliers provide a better trade-off between accuracy and performance by offering 17 design points that lie on the Pareto surface. As can be observed that most of our non-dominated design points have lower average absolute relative error values than those offered by the non-dominated design points of EvoApprox. Further, most of the dominated design points of EvoApprox have a higher PDP × LUTs value than our proposed designs. Due to the higher error profile and lower performance of the *Approx-1* design than *Approx-2* and *Approx-3* designs, none of the non-dominated design points hosts *Approx-1* as a sub-multiplier. Figure 4.14 also compares the exclusive hypervolume contribution of our proposed non-dominated designs points with the Pareto design points of the EvoApprox library. Due to lower PDP × LUTs and average absolute relative error values of our non-dominated design points, they have more significance than the EvoApprox design points.

4.6.2.4 Quality Evaluation of Approximate Multipliers for Application Kernels

As discussed previously in Sect. 4.6.2.2, an approximate operator can still produce acceptable quality application-level results despite having a high error profile. To evaluate the application-level impact of the proposed 8 × 8 unsigned approximate multipliers, we have used them for the multiply-mode-based image blending application and SUSAN application-based image smoothing accelerator. For the image blending filter, we have utilized our approximate multipliers in the Python-based behavioral model of the application and used it for ten random test images from USC-SIPI database [20]. Table 4.7 shows the average PSNR values for different approximate multiplier-based filters compared to the accurate multiplier-based filter. Despite the high error profile of *Approx-1*-based multipliers, they produce images with acceptable PSNR.[9] Figure 4.15 presents the visual output along with respective PSNR values for a single image.

---

[9] The PSNR value ranges from 30 to 50 dB for 8-bit data representation in image and video processing [21].

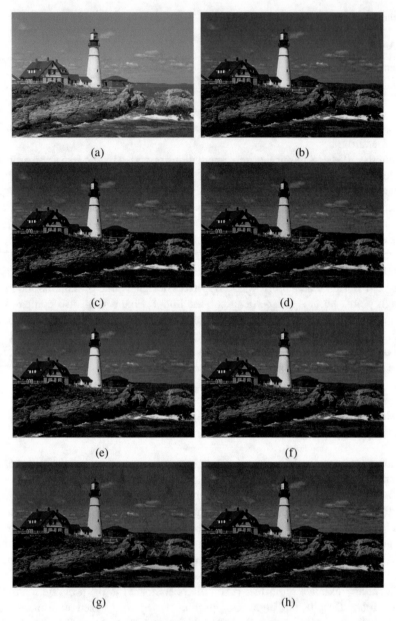

Fig. 4.15 Image blending application (multiply mode); the PSNR values of the approximate multiplier-based filters are computed with respect to the accurate multiplier-based filter:*Approx-1*-a PSNR = 37.90, and *Approx-1*-p PSNR = 30.17, *Approx-2*-a PSNR = 52.20 dB, *Approx-2*-p PSNR = 30.85 dB, *Approx-3*-a PSNR = 33.37, *Approx-3*-p PSNR = 29.50. (**a**) Original image. (**b**) Accurate. (**c**) *Approx-1*-a. (**d**) *Approx-1*-p. (**e**) *Approx-2*-a. (**f**) *Approx-2*-p. (**g**) *Approx-3*-a. (**h**) *Approx-3*-p

Table 4.8 PSNR values of 8×8 approximate multipliers

Multiplier architecture	SUSAN accelerator PSNR [dB]
Accurate	∞
Approx-2-a	33.716
Approx-2-p	25.602
S1	47.493
S2	17.944
Approx-2-a$_s$ (*Approx-2*-a swapped inputs)	59.119
Approx-2-p$_s$ (*Approx-2*-p swapped inputs)	27.366

We have also synthesized the SUSAN application-based image smoothing accelerator with Vivado's multiplier IP and our proposed *Approx-2*-a and *Approx-2*-p multipliers using Xilinx Vivado. Our approximations offer 17% and 17.2% gains in LUT utilization for *Approx-2*-a and *Approx-2*-p multipliers, respectively, with insignificant output quality loss. Table 4.8 presents the PSNR values of the SUSAN image smoothing accelerator, using accurate multiplier, *Approx-2*-based multipliers, and state-of-the-art multipliers S1 [2] and S2 [3], respectively. The results show that *Approx-2*-based multipliers provide better PSNR values than those displayed by the S2 multiplier. The approximate multiplier S1, apparently, produces a better PSNR value than those produced by *Approx-2*-a and *Approx-2*-p. However, exploiting the asymmetric nature of *Approx-2* multiplier, the mutual swapping of all input values to our approximate multipliers for SUSAN image smoothing accelerator and input image under consideration results in enhanced output qualities with higher PSNR values, as shown in Table 4.8. Hence, depending upon the input data and the application under analysis, *Approx-2*-a, *Approx-2*-p, or *Approx-2*-a$_s$, *Approx-2*-p$_s$ can be deployed for achieving enhanced output accuracy.

4.6.3 Evaluation of the Proposed Approximate Signed Multiplier

We have compared the proposed approximate signed multiplier—*Booth-Approx*—with the implementations presented in S1 [2], S2 [3], EvoApprox signed multipliers [14], and precision-reduced P(N,2) multiplier. As discussed in the previous section, P(N,2) denotes truncating the two LSBs of each operand and utilizing an N-2 × N-2 multiplier for the multiplication. For the unsigned multipliers S1 and S2, we have shown implementation results with/without implementing signed-unsigned converters. As described in Chap. 3, these converters provide $2's$ complement signed numbers to the unsigned multipliers. Further, we have also analyzed the output accuracy of *Booth-Approx* in comparison to the fixed-width multipliers S3 [5], S4 [6], S5 [7], and S6 [8]. The accuracy analysis of all presented multipliers has been computed using multiple uniform distributions of all input combinations

Table 4.9 Error analysis of 8 × 8 approximate signed multipliers

Design	Error occurrences %	Maximum error	Average abs. error	Max. abs. relative error	Avg. abs. relative error
Booth-Approx	90.56	361	85.01	6	0.091
S1 [2]	86.46	7225	1842.44	1	0.362
S2 [3]	34.19	882	118.875	1	0.0223
P(8,2)	93	759	149.78	15	0.121

(unless stated otherwise). Moreover, the C and Python-based behavioral model of the proposed *Booth-Approx* is also deployed in the Gaussian image smoothing application and an ANN for testing the effects of *Booth-Approx* in the real-world applications.

4.6.3.1   Error Analysis of *Booth-Approx*

Table 4.9 presents the error analysis of *Booth-Approx* multiplier along with precision-reduced P(8,2) and other state-of-the-art approximate multipliers (using signed-unsigned converters). Since the number "−128" cannot be represented using sign-magnitude format for 8 × 8 multipliers with 8-bit operands, the observed *maximum error* magnitude in S1 and S2 is 16384. However, to show a fair comparison, the 8-bit operands' range for computing the *maximum error* is limited to [−127, +127] for designs in S1 and S2. As shown by the highlighted cells in the table, the *Booth-Approx* has the least maximum error magnitude and average absolute error among all presented multipliers. Further, it can be observed that *Booth-Approx* is better than the P(8,2) multiplier across all the presented error parameters.

To explore the error occurrences of *Booth-Approx* multiplier, Fig. 4.16 presents the probabilistic error analysis for an 8 × 8 *Booth-Approx* multiplier. These results have been obtained for a uniform distribution of all input combinations. As shown by the bit inaccuracy histograms in Fig. 4.16a, the probability of errors in individual product bits reduces for higher-order product bits. The probability mass functions (PMF) of errors, depicted in Fig. 4.16b, also show that the majority of occurred errors have small values. This is also verified by the relative error distribution plot shown in Fig. 4.16c. As shown by the results, most of the final products have very small relative errors (on average less than 0.1). This behavior is in accordance with our design modifications discussed in Sect. 4.5. Since higher-order multipliers have long carry chains, the errors generated by incorrect input carries diminish for higher-order product bits. Moreover, a constant "0" multiplicand/multiplier results in an accurate "0" result.

Further, Fig. 4.17 shows the error metrics of our proposed design compared to that of truncated and truthfully rounded multipliers. The maximum, average, and mean squared errors of each design are normalized with 2^n, 2^n, and 2^{2n},

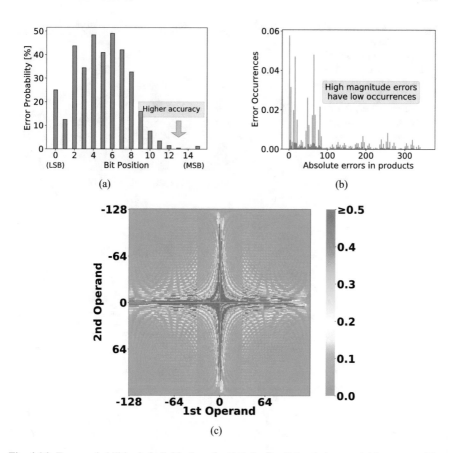

Fig. 4.16 Error probabilities in individual product bits for 8 × 8 *Booth-Approx*. (**a**) Inaccuracy bit histogram. (**b**) PMFs of error values. (**c**) Relative error distribution

respectively—*n* being the input bit width [7]. As seen in the figure, our proposed design outperforms the other designs for normalized maximum error for larger designs, such as 12 × 12 and 14 × 14. The normalized average error of *Booth-Approx* is also always less than S3 and S4 designs, as shown in Fig. 4.17b. Further, as shown in Fig. 4.17c, the normalized mean square error of *Booth-Approx* is always less than S6. It should be noted that the faithful rounding, after truncation, usually involves some form of compensation to reduce the error [22]. This additional compensation logic is optimized for ASIC-based implementation and can result in large overheads in FPGAs. For instance, the implementation of the compensation logic used in [22] results in 90 LUTs being used for a 8 × 8 multiplier.

Fig. 4.17 Comparison with truncated and rounded multipliers: S3 [5], S4 [6], S5 [7], and S6 [8]. (**a**) Normalized maximum error (ne_{max}). (**b**) Normalized average error (ne_{avg}). (**c**) Normalized mean square error (ne_{mse})

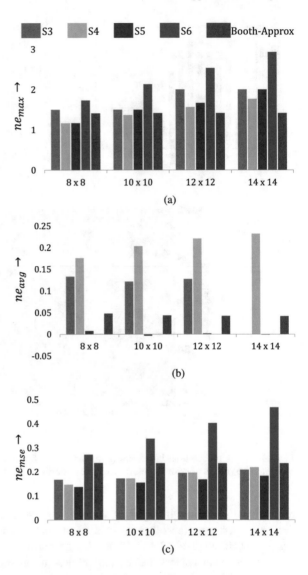

4.6.3.2   Performance Characterization of *Booth-Approx*

Tables 4.10 and 4.11 show the comparison of the resource utilization, critical path delay, and energy consumption of the proposed *Booth-Approx* multiplier with different state-of-the-art accurate and approximate multipliers. As the *Booth-Approx* design is based on the *Booth-Opt* design (presented in Chap. 3), the tables also include the implementation results of the *Booth-Opt* design. The results in Tables 4.10 and 4.11 also incorporate the signed-unsigned converters for the unsigned multipliers S1 and S2. As shown by the results, the proposed

Table 4.10 Implementation results of different multipliers. The S1 and S2 are implemented with the signed-unsigned converters. The CPD and PDP are in ns and pJ, respectively. The highlighted cells show the minimum value for the respective performance metric across all the designs

Design	4 × 4			6 × 6			8 × 8		
	LUTs	CPD	PDP	LUTs	CPD	PDP	LUTs	CPD	PDP
Booth-Opt	12	2.15	1.09	24	3.09	2.67	40	4.25	5.14
Booth-Approx	11	1.94	0.81	22	2.64	2.18	37	3.41	4.22
S1 [2]	18	2.23	0.86	49	4.82	3.78	92	4.99	7.10
S2 [3]	20	2.12	0.87	52	4.83	4.91	86	4.89	7.42
P(N,2)	2	0.66	0.06	23	1.58	1.21	43	2.15	3.06
Vivado speed [15]	18	2.14	1.06	41	3.43	3.26	74	3.54	5.73
Vivado area [15]	30	2.91	2.25	47	3.39	4.73	88	3.45	9.07

Table 4.11 Implementation results of different multipliers. The S1 and S2 are implemented with the signed-unsigned converters. The CPD and PDP are in ns and pJ, respectively. The highlighted cells show the minimum value for the respective performance metric across all the designs

Design	12 × 12			16 × 16			24 × 24		
	LUTs	CPD	PDP	LUTs	CPD	PDP	LUTs	CPD	PDP
Booth-Opt	84	6.31	11.96	144	7.64	21.15	312	11.37	49.17
Booth-Approx	79	5.30	10.17	137	6.88	19.14	301	10.99	48.26
S1 [2]	228	6.98	20.80	404	7.03	22.32	895	9.43	101.63
S2 [3]	189	6.37	20.77	330	6.59	20.39	777	9.45	97.48
P(N,2)	102	3.52	8.97	214	4.11	14.76	514	6.07	53.97
Vivado speed [15]	162	4.20	19.79	286	4.27	34.35	627	5.98	77.25
Vivado area [15]	175	5.00	15.33	326	5.04	35.25	592	5.55	78.41

Booth-Approx always requires fewer LUTs than other state-of-the-art accurate and approximate multipliers for different sizes of multipliers. The reductions in total utilized LUTs with respect to Vivado's area-optimized multiplier IPs vary between 49.1% (for 24 × 24) and 63.3% (for 4 × 4). For example, compared to the 8 × 8 multiplier IP, the proposed *Booth-Approx* shows 57.9% reduction in the total utilized LUTs. Only for smaller multipliers, such as 4 × 4, the P(N,2) occupies fewer LUTs than the *Booth-Approx*. The reason for this is because a P(4,2) multiplier is implemented using a 2 × 2 multiplier, and only two LUTs are sufficient to implement a 2 × 2 multiplier by exploiting the two outputs (O5 and O6) of a 6-input LUT.

Compared to the various state-of-the-art approximate multipliers and Vivado's area-/speed-optimized IPs, the *Booth-Approx* offers comparable critical path delays, as shown in Tables 4.10 and 4.11. The slight increase in the critical path delays of the proposed multiplier is due to the sequential computation of booth-encoded partial products, as discussed in Sect. 4.5. However, for smaller multipliers, such as 4 × 4, 6 × 6, and 8 × 8, *Booth-Approx* has a lower CPD than approximate S1 and S2 designs and Vivado area-/speed-optimized multiplier IPs. For example, compared to the 6 × 6 area-optimized IP, *Booth-Approx* offers a 22% reduction in the critical path delay.

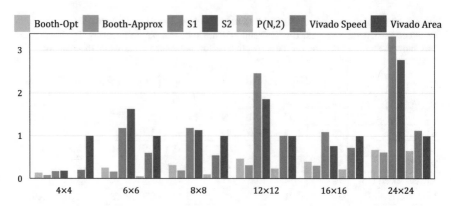

Fig. 4.18 Product of normalized performance metrics. Values are normalized to Vivado area-optimized multiplier IP. A smaller value reflects better performance

Table 4.12 Comparison of implementation results of *Booth-Approx* multipliers with unsigned approximate multipliers

Design	8 × 8			16 × 16		
	LUTs	CPD [ns]	PDP [pJ]	LUTs	CPD [ns]	PDP [pJ]
Booth-Approx	37	3.41	4.22	137	6.88	19.14
S1 [2]	57	3.13	4.70	377	4.57	24.03
S [3]	80	2.33	5.13	294	4.06	21.73

Booth-Approx is also more energy-efficient than state-of-the-art approximate designs and Vivado area-/speed-optimized IPs for different sizes of multipliers. Compared to the Vivado area-optimized IP, *Booth-Approx* offers up to a 63.9% reduction in the PDP value. For example, compared to the 8 × 8 area- and speed-optimized IPs, *Booth-Approx* provides 53.4% and 26.2% reduction in the PDP value, respectively.

To highlight the efficacy of *Booth-Approx* multiplier, Fig. 4.18 shows the product of normalized values of total utilized LUTs, CPD, and PDP for each design across different bit widths. All values have been normalized to the corresponding values of Vivado area-optimized multiplier IP. A smaller value of the product (LUTs × CPD × PDP) presents an implementation with a better performance. Although for smaller designs, the P(N,2) multiplier performs better than *Booth-Approx*, the performance gains do not scale proportionally for higher-order P(N,2) multipliers. For example, in 24 × 24 multipliers, *Booth-Approx* provides a 5.2% reduction in the product of the normalized performance metrics compared to the P(N,2) multiplier. Moreover, the error analysis of the approximate multipliers, presented in the previous subsection, shows the lower accuracy of P(N,2) multipliers across all error metrics.

We have also compared *Booth-Approx* with the state-of-the-art *unsigned* designs S1 and S2 without using *signed-unsigned converters* for them. As shown by the results in Table 4.12, our proposed approximate *signed* designs still offer better LUTs reductions than all other implementations. The unsigned implementations

Fig. 4.19 Performance
comparison of 8 × 8
Booth-Approx with signed
multipliers of EvoApprox.
The PDP is in pJ

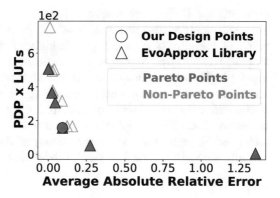

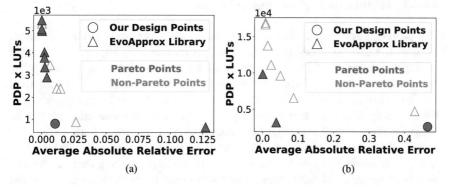

Fig. 4.20 Performance comparison of *Booth-Approx* with signed multipliers of EvoApprox. The
PDP is in pJ. (**a**) 12 × 12. (**b**) 16 × 16

have slightly reduced critical path delays than the proposed multiplier. However,
the energy consumption of *Booth-Approx* is still better than state-of-the-art designs.
For example, *Booth-Approx* offers 20% reduction in energy consumption when
compared with the 16×16 unsigned multiplier S1.

Finally, to provide a more exhaustive comparison of *Booth-Approx* with state-
of-the-art designs, Figs. 4.19 and 4.20 compare *Booth-Approx* with the signed
multipliers provided by the EvoApprox library [14]. The EvoApprox signed
multipliers library hosts 13, 14, and 9 accurate and approximate 8 × 8, 12 × 12,
and 16 × 16 multipliers, respectively. We have used the PDP × LUTs as the
performance metric and the average absolute relative error as the accuracy metric to
compare all the designs. As shown in Figs. 4.19 and 4.20, the *Booth-Approx* design
always lies on the Pareto front for different sizes of multipliers. *Booth-Approx* has
lower values of PDP × LUTs and average absolute relative error than most of the
EvoApprox designs. For example, Fig. 4.19 shows that two EvoApprox designs have
lower PDP × LUTs values than the *Booth-Approx*. However, these designs do not
compute any logic for the lower six product bits and utilize constant "0" for these
product bits. These designs approximately compute the remaining product bits by

deploying specific Boolean functions. However, both of these designs have higher average absolute relative error than *Booth-Approx*. One of the non-dominated design points of EvoApprox is an accurate multiplier implementation having 0 average absolute relative error. Similar results are also obtained for 12×12 multipliers in Fig. 4.20a. The non-dominated EvoApprox multiplier having a lower PDP $\times$ LUTs value than *Booth-Approx* truncates the least significant six product bits. For 16×16 multipliers, only two designs of EvoApprox are non-dominated. One of these designs is an accurate multiplier, and the other design is a precision-reduced P(16,2) multiplier. These results validate the efficacy of *Booth-Approx* in providing better performance and accuracy than state-of-the-art designs.

4.6.3.3 High-Level Application Testing

The proposed *Booth-Opt* and *Booth-Approx* were used in the implementation of finite impulse response (FIR) filter for image processing applications. Typical hardware realization of FIR filters involves the implementation of N multipliers and N adders—N being the number of *taps* in the filter—to implement convolution. For this experiment, we have used *Gaussian smoothing* as a test case for evaluating the efficiency of using the proposed signed multipliers. Gaussian smoothing of an image involves two-dimensional (2D) convolution of the image with a *Gaussian kernel*. This 2D convolution can also be achieved by a two-stage (2S) method that entails successive one-dimensional (1D) convolutions along each of the two directions—horizontal and vertical. These two methods can have large differences in the resource utilization of their realizations—$O(n+m)$ and $O(nm)$ for a window size of $n \times m$ for 2S and 2D, respectively.

We performed experiments for convolution window sizes of 3×3, 5×5 and 7×7 and compared the effects of using accurate and approximate signed multipliers (8×8) on the resource utilization and the degradation in processed image quality. We have used the resource utilization—in terms of $LUTs$ used for the multipliers only—of Vivado's area-optimized multiplier IP-based implementation and the corresponding output image quality as the baseline for comparison. Two metrics were used for processed image quality, (1) PSNR, an estimation of the noise component in the image, and (2) Structural Similarity Index (SSIM), a measure of the structural similarity between the two images.

Table 4.13 shows the comparison results for Gaussian smoothing using *all-accurate* (*Booth-Opt* multipliers only) and *all-approximate* (*Booth-Approx* multipliers only) FIR filters. The table data denotes the average values from processing 15 *miscellaneous* images in USC-SIPI database [20]. Area reduction estimates are similar to those presented in Sect. 4.6.3. The PSNR and SSIM values correspond to the comparison of processed images from the *all-accurate* and *all-approximate* implementations. Average PSNR of up to 52.36 and 47.55 were observed for 2D and 2S modes, respectively. Similarly, SSIM of up to 0.99 and 0.97 were observed, respectively, for the two modes using the approximate multipliers.

Table 4.13 Comparison of *all − accurate* and *all − approximate* multipliers-based FIR filters

Convolution window size window size	Relative LUTs reduction (w.r.t. Vivado's IP in %)		PSNR		SSIM	
	Booth-Opt	Booth-Approx	2D	2S	2D	2S
3 × 3	54.6	57.95	50.50	45.72	0.98	0.96
5 × 5	54.6	57.95	51.85	45.19	0.99	0.97
7 × 7	54.6	57.95	52.36	47.55	0.99	0.95

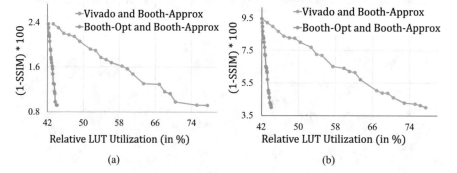

Fig. 4.21 Pareto fronts for combinations of accurate and approximate multipliers in an FIR filter with 49 *taps* for processing of two benchmark images—(**a**) Lena (**b**) Cameraman. The metrics are relative to that of an all-accurate Vivado multiplier-based design

The *all-accurate* and *all-approximate* implementations denote the two extremes of the possible multiplier configurations in an FIR filter implementation. We have used Genetic Algorithm (GA)-based multi-objective design space exploration (DSE) for finding an appropriate combination of multiplier types with a trade-off between resource utilization and processed image quality. An *individual* of the *population* is denoted by a sequence string specifying the type of multiplier used at each position of the FIR filter's hardware implementation. Starting population of 100 individuals and a maximum of 25 generations have been used for the DSE. Two-point crossover with a probability of 0.5 and a mutation probability of 0.1 are used for *evolution*. We have used two sets of DSE-related experiments. In the first case, the multiplier types are allowed to vary among either Vivado's signed multiplier IP type or our proposed *Booth-Approx* type. Similarly, the choices in the second experiment are restricted to both the *Booth-Opt* and *Booth-Approx*. Each possible combination of the choices for the type of multiplier—Vivado's IP-based, *Booth-Opt*, or *Booth-Approx*—used in the FIR filter's hardware implementation results in a unique point on the design space. Figure 4.21a and b shows the Pareto fronts generated from the two DSE experiments—Vivado + *Booth-Approx* and *Booth-Opt* + *Booth-Approx* for the images used in Fig. 4.22a and b, respectively. The results correspond to the 2D processing mode with a 7×7-sized convolution window. As evident from the figure, using our proposed multiplier architectures results in a large reduction in resource utilization while still providing an equivalent number of Pareto

Fig. 4.22 Comparison of Gaussian smoothing of images with different configuration of multiplier approximation. Processing mode: two-stage processing, 7×7 window size and 8×8 multiplier

front design points. Therefore, using our proposed multipliers can result in better application-level optimization.

The extent of application-level improvements depends upon the number of operators in the design that can be approximated. We performed experiments for estimating the effect of the proposed approximate design on ANNs. As a case study, we implemented the proposed multiplier designs for a single fully connected layer of an ANN with 85 physical artificial neurons. The resulting design with 8×8 *Booth-Approx* multipliers resulted in significantly lesser resource utilization (6825 $CLBs$ compared to 8285 $CLBs$), critical path delay (3.976 ns compared to 4.423 ns), and energy (PDP of 10.88 nJ compared to 12.31 nJ) than one using *Booth-Opt*. On the other hand, the behavioral analysis of a sample ANN-based inference of MNIST Fashion database [23] showed only 0.04% reduction in accuracy due to the usage of approximate multipliers. Hence, in such applications, the resources saved using the approximate-based design can be used to instantiate more artificial neurons to provide improved parallelism without any significant reduction in inference accuracy. Therefore, the small improvements for a single approximate multiplier can result in significant overall improvements for similar larger applications.

4.7 Conclusion

In this chapter, we presented various designs of unsigned and signed approximate multipliers optimized for FPGAs. Most of these designs are based on the accurate multipliers presented in Chap. 3. The main design target of these designs is to intelligently waive the output accuracy of a multiplier to improve the corresponding performance, i.e., resource utilization, critical path delay, and energy consumption gains. Towards this end, we presented three 4×4 approximate unsigned multiplier designs offering different output accuracy and performance. Overall the proposed *Approx-2*-based designs offer higher output accuracy than other proposed and state-of-the-art approximate designs. Similarly, the proposed *Approx-3*-based designs offer better LUT utilization, lower CPD, and reduced energy than other proposed approximate and state-of-the-art designs. We have also proposed an approximate ternary adder for implementing higher-order approximate multipliers from the building 4×4 multipliers. The proposed approximate ternary adder eliminates the propagation of carries and offers lower critical path delay than the corresponding accurate ternary adders. For example, compared to Vivado's 8×8 area-optimized unsigned multiplier IP, the *Approx-3*-based 8×8 multiplier using approximate ternary adder offers 38.9%, 56.9%, and 67.3% reduction in the total utilized LUTs, CPD, and PDP values. This chapter has also presented a radix-4 Booth's multiplication algorithm-based approximate signed multiplier referred to as *Booth-Approx*. Compared to the unsigned multipliers, we have considered array-based implementation for the proposed *Booth-Approx* multiplier. *Booth-Approx* provides better LUT utilization than state-of-the-art accurate and approximate designs. For example, compared to the 8×8 Vivado's area-optimized multiplier IP, Booth-Approx offers a 57.9% reduction in the total utilized LUTs. We have also evaluated the efficacy of our proposed approximate multipliers in various applications.

References

1. K. He, X. Zhang, S. Ren, J. Sun, Deep residual learning for image recognition, in *Proceedings of the IEEE Conference on Computer Vision and Pattern Recognition* (2016), pp. 770–778
2. S. Rehman, W. El-Harouni, M. Shafique, A. Kumar, J. Henkel, J. Henkel, Architectural-space exploration of approximate multipliers, in *2016 IEEE/ACMInternational Conference on Computer-Aided Design (ICCAD)* (IEEE, Piscataway, 2016), pp. 1–8
3. P. Kulkarni, P. Gupta, M. Ercegovac, Trading accuracy for power with an underdesigned multiplier architecture, in *2011 24th Internatioal Conference on VLSI Design* (IEEE, Piscataway, 2011), pp. 346–351
4. W. Liu, L. Qian, C. Wang, H. Jiang, J. Han, F. Lombardi, Design of approximate radix-4 booth multipliers for error-tolerant computing. IEEE Trans. Comput. **66**(8), 1435–1441 (2017)
5. K.-J. Cho, K.-C. Lee, J.-G. Chung, K.K. Parhi, Design of low-error fixed-width modified booth multiplier IEEE Trans. Very Large Scale Integr. Syst. **12**(5), 522–531 (2004)
6. M.-A. Song, L.-D. Van, S.-Y. Kuo, Adaptive low-error fixed-width Booth multipliers.IEICE Trans. Fund. Electron. Commun. Comput. Sci. **90**(6), 1180–1187 (2007)

7. J.-P. Wang, S.-R. Kuang, S.-C. Liang, High-accuracy fixed-width modified Booth multipliers for lossy applications. IEEE Trans. Very Large Scale Integr. Syst. **19**(1), 52–60 (2009)

8. S.-J. Jou, M.-H. Tsai, Y.-L. Tsao, Low-error reduced-width Booth multipliers for DSP applications. IEEE Trans. Circuits Syst. I: Fund. Theory Appl. **50**(11), 1470–1474 (2003)

9. T. Yang, T. Ukezono, T. Sato, A low-power highspeed accuracy-controllable approximate multiplier design, in *2018 23rd Asia and South Pacific Design Automation Conference (ASP-DAC)* (IEEE, Piscataway, 2018), pp. 605–610

10. N. Van Toan, J.-G. Lee, FPGA-based multi-level approximate multipliers for high-performance error-resilient applications. IEEE Access **8**, 25481–25497 (2020)

11. Y. Guo, H. Sun, S. Kimura, Small-area and low-power FPGA-based multipliers using approximate elementary modules, in *2020 25th Asia and South Pacific Design Automation Conference (ASP-DAC)* (IEEE, Piscataway, 2020), pp. 599–604

12. H. Waris, C. Wang, W. Liu, F. Lombardi, AxBMs: approximate radix-8 booth multipliers for high-performance FPGA-based accelerators. IEEE Trans. Circuits Syst. II: Exp. Briefs **68**(5), 1566–1570 (2021)

13. B.S. Prabakaran, V. Mrazek, Z. Vasicek, L. Sekanina, M. Shafique, ApproxFPGAs: embracing ASIC-based approximate arithmetic components for FPGA-based systems, in *2020 57th ACM/IEEE Design Automation Conference (DAC)* (IEEE, Piscataway, 2020), pp. 1–6

14. V. Mrazek, L. Sekanina, Z. Vasicek, Libraries of approximate circuits: automated design and application in CNN accelerators. IEEE J. Emerg. Sel. Topics Circuits Syst. **10**(4), 406–418 (2020)

15. Xilinx LogiCORE IP v12.0 (2015). https://www.xilinx.com/support/documentation/ipdocumentation/multgen/v120/pg108-multgen.pdf

16. N. Brunie, F. de Dinechin, M. Istoan, G. Sergent, K. Illyes, B. Popa, Arithmetic core generation using bit heaps, in *2013 23rd International Conference on Field Programmable Logic and Applications* (2013), pp. 1–8

17. A.K. Verma, P. Brisk, P. Ienne, Variable latency speculative addition: a new paradigm for arithmetic circuit design, in *2008 Design, Automation and Test in Europe* (2008), pp. 1250–1255

18. V. Mrazek, R. Hrbacek, Z. Vasicek, L. Sekanina, EvoApprox8b: library of approximate adders and multipliers for circuit design and benchmarking of approximation methods, in *Design, Automation Test in Europe Conference Exhibition (DATE), 2017* (2017), pp. 258–261

19. E. Zitzler, D. Brockhoff, L. Thiele, The hypervolume indicator revisited: on the design of Pareto-compliant indicators via weighted integration, in *International Conference on Evolutionary Multi-Criterion Optimization* (Springer, Berlin, 2007), pp. 862–876

20. SIPI Image Database (2019). http://sipi.usc.edu/database/database.php?volume=misc

21. U. Sara, M. Akter, M.S. Uddin, Image quality assessment through FSIM, SSIM, MSE and PSNR—a comparative study. J. Comput. Commun. **7**(3), 8–18 (2019)

22. H.-J. Ko, S.-F. Hsiao, Design and application of faithfully rounded and truncated multipliers with combined deletion, reduction, truncation, and rounding. IEEE Trans. Circuits Syst. II: Exp. Briefs **58**(5), 304–308 (2011)

23. H. Xiao, K. Rasul, R. Vollgraf, Fashion-mnist: a novel image dataset for benchmarking machine learning algorithms (2017). arXiv preprint arXiv:1708.07747

Chapter 5
Designing Application-Specific Approximate Operators

5.1 Introduction

The paradigm of approximate computing has emerged as a viable solution for implementing high-performance and energy-efficient hardware accelerators for error-tolerant applications [1]. The principles of approximate computing can be applied at any layer of the computation stack [2]. Among these layers of approximation, circuit-level techniques have been a major focus of research for resource-constrained embedded systems [3–14]. However, the state-of-the-art approximate arithmetic blocks have two main limitations:

(1) These approximate blocks lack a consistent design methodology and have considered different strategies for introducing approximations to obtain performance gains. For example, the 4×4 unsigned approximate multiplier *Approx-1* and the signed approximate multiplier *Booth-Approx*, presented in Chap. 4, had different design methodologies. Further, for a fixed n-bit operator, most of the state-of-the-art designs provide a fixed output accuracy and corresponding performance gain. For example, compared to a 4×4 accurate multiplier, *Approx-1* and *Booth-Approx* provide a 25% and 63% reduction in the total utilized LUTs, respectively, and 0.072 and 0.55 degradation in the output quality (average relative error), respectively. Therefore, to design a new approximate block with a modified accuracy-performance constraint, these techniques may not be sufficient and call for exploring some other approximation techniques. The modular design methodology, as used in [13] and [12], of designing N × N approximate multipliers from $\frac{N}{2}$ approximate multipliers provides a limited design space and may not be sufficient to satisfy the modified accuracy-performance constraint.

(2) Most of the state-of-the-art approximate arithmetic operators are designed without considering an application's accuracy-performance constraints. The application agnostic-design methodology can result in approximate operators

© The Author(s), under exclusive license to Springer Nature Switzerland AG 2023
S. Ullah, A. Kumar, *Approximate Arithmetic Circuit Architectures for FPGA-based Systems*, https://doi.org/10.1007/978-3-031-21294-9_5

which may not satisfy an application's accuracy-performance constraints. For example, Fig. 5.1 presents the performance-accuracy analysis for two different applications using the approximate signed multipliers library EvoApprox from [6, 14]. We have used the PDP × LUTs metric for reporting the corresponding performance of each application by implementing it on Xilinx UltraScale FPGA using Xilinx Vivado. For PDP, the dynamic power is computed in μW, and the critical path delay is reported in ns. Similarly, we have used Python-based models to report each application's output accuracy for various approximate multipliers. These results identify the approximate multipliers which contribute to non-dominated hardware accelerators for each application. Figure 5.1a describes the hardware accelerator results for the QRS peak detection in Electrocardiographic (ECG) signals. To report the application's output accuracy for various approximate multipliers, we have used the *accuracy of the peak detection* metric. The results show that *nine* different approximate multipliers (out of 13) contribute to non-dominated design points. As shown in Fig. 5.1b, the application-agnostic design of the multipliers has resulted in only *two* non-dominated design points (accelerator designs) of the Gaussian image smoothing filter (for 45 images). Moreover, the application-level analysis also reveals that the accurate multiplier *1KV8* from the EvoApprox library [14] does not contribute to any non-dominated accelerator design point for any application.

Therefore, it is necessary to define a generic and optimized design methodology that can generate application-specific approximate operators satisfying its accuracy-performance constraints. Towards this end, we present *AppAxO*: a methodology for designing *app*lication-specific *a*pproximate arithmetic *o*perators for FPGAs-based systems in this work. As shown in Fig. 5.2, the proposed methodology involves

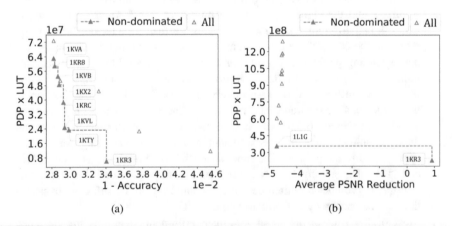

(a)   (b)

Fig. 5.1 Application-level performance-accuracy analysis of utilizing approximate multipliers from [14] for three different applications (**a**) ECG QRS peak detection (**b**) Gaussian image smoothing filter

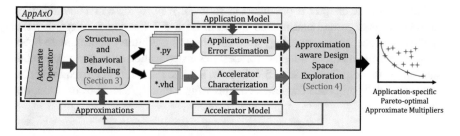

Fig. 5.2 Proposed framework for *AppAxO*

circuit-level modeling and novel DSE methods for fast design of approximate arithmetic operators that can leverage the inherent robustness of error-tolerant applications. We have considered multipliers and adders as example operators to discuss our methodology and present our results; however, the proposed methodology is generic and can be used for designing any *soft logic-based* operators including dividers. Our proposed implementations utilize the 6-input LUT and associated carry chains of modern FPGAs building blocks. Our novel contributions include:

5.1.1 Contributions

- *A systematic methodology for approximate operators generation:* We provide a systematic and generic methodology for implementing approximate operators of arbitrary size for FPGA-based systems. Our methodology utilizes the 6-input LUTs and the associated carry chains of FPGAs to implement approximate operators according to input configuration. For instance, the input configurations—a binary string—identify the LUTs, in an accurate multiplier implementation, that should be disabled to realize an approximate multiplier. For an M × N accurate multiplier, utilizing 'K' LUTs, our methodology provides 2^K approximate multipliers with different accuracy and performance parameters.
- *Application-specific multiplier configurations:* We utilize a MBO-based exploration method to generate only those approximate multiplier configurations (a binary string) that satisfy an application's accuracy and performance constraints. Our proposed multiplier generation methodology uses these configurations to implement the respective multipliers for the application.
- *An efficient DSE methodology:* We utilize various ML models to propose a GA-based DSE methodology. Our methodology deploys various ML models to explore the large design space of individual multipliers and their utilization in various applications by estimating the behavioral accuracy and corresponding performance gains.

The rest of the chapter is organized as follows. We provide a brief overview of the relevant background and related works in Sect. 5.2. Section 5.3 describes the systematic modeling methodology used for designing arbitrary approximate operators. The DSE methods adopted for designing application-specific approximate multipliers are described in Sect. 5.4. In Sect. 5.5, the results from the experimental evaluation of the proposed framework are presented. Finally, Sect. 5.6 concludes the chapter.

5.2   Related Work

The work presented in [10] has used a Cartesian Genetic Programming (CGP)-based technique to design unsigned approximate multipliers for ANNs. This technique utilizes a $2D$ array of 2-input logic gates to represent multipliers. The initial population for their technique consists of *three* accurate and a few approximate unsigned multipliers. In each iteration of the CGP, a new set of multipliers is generated according to a predefined approximation error ϵ. The multipliers' efficacy is assessed by deploying these multipliers in an ANN and evaluating the network's output accuracy after retraining. Depending upon the network's output accuracy, the approximation error value ϵ can be adjusted for the next iteration of the CGP. However, this technique does not consider multiplier performance parameters, such as critical path delay and dynamic power, while generating new approximate multipliers. The output accuracy of the network is the only factor deciding the generation of an approximate multiplier. Further, the generated approximate multipliers are unsigned, and separate circuitry for calculating the product sign has been used to deploy ANNs.

Some recent works have also utilized various machine learning techniques for performing application-specific DSE on existing approximate arithmetic operators. For example, the authors of [15] have used various ML models to perform an efficient DSE for Sobel filter considering approximate adders. In their proposed technique, the non-feasible approximate adders (from an existing library of approximate adders) are filtered out by computing the weighted mean error distance and hardware performance metrics of each approximate circuit. The feasible approximate adders are then used to train machine learning models to estimate the behavioral accuracy and performance parameters of the Sobel filter accelerator. In Chap. 4, we have also used GA-based multi-objective design space exploration for Gaussian smoothing filter. Utilizing a set of an accurate and an approximate multipliers (*Booth-Opt*, *Booth-Approx*), we have used GA to find a feasible combination of multiplier types with a trade-off between output quality and LUT utilization. The authors of [12] have also utilized a depth-first search-based methodology to perform architectural space exploration for designing larger multipliers from 2×2 approximate multipliers. Their technique utilizes a weighted average of the area and power to identify feasible design points.

Applications from different domains exhibit diverse error tolerance for approximate arithmetic operators and hold different output accuracy and performance (resources, latency, and power) requirements. However, the various related state-of-the-art works, summarized in Table 5.1, do not offer a methodology for generating approximate operators according to an application's accuracy and performance constraints. To address the limitations of the state-of-the-art works, we present *AppAxO* methodology to generate new approximate arithmetic operators according to an application's accuracy and performance constraints. Towards this end, *AppAxO* utilizes various machine learning models to identify feasible approximate operators for an input application efficiently. Our proposed methodology is generic and can generate approximate circuits for any arithmetic operator which utilizes LUTs and carry chains for its implementation. It should be noted that *AppAxO* focuses on circuit-level approximations only. Other optimization techniques, such as exploring the impact of High-level Synthesis (HLS) directives and approximations across multiple layers of computation stack, are orthogonal to our proposed work. These optimization techniques can be integrated with our proposed methodology.

5.3 Modeling Approximate Arithmetic Operators

AppAxO proposes a systematic methodology for designing approximate arithmetic operators from the corresponding accurate implementations of the operators. Specifically, we enable better utilization of the FPGA resources (LUTs and carry chains), resulting in higher packing efficiency (CLB utilization) than state-of-the-art works [16].

5.3.1 Accurate Multiplier Design

We have used Baugh-Wooley's multiplication algorithm in this chapter to implement a signed multiplier for our *AppAxO* methodology. However, the proposed methodology for approximation is equally applicable to other multiplication algorithms, such as Booth's algorithm. Please note that Chap. 3 has used Baugh-Wooley's and Booth's multiplication algorithms to present accurate signed multipliers. Moreover, Baugh-Wooley's multiplication algorithm is also described in the *Preliminaries* in Chap. 2. However, for better readability, we show the multiplication algorithm again in Eq. 5.1 and the corresponding LUT configurations in Fig. 5.3. Utilizing the LUT configurations, Fig. 5.4 presents an accurate signed 4×4 multiplier as an example to explain the approximation methodology in the next section. The ① in the figure shows the addition of 1s at 2^{N-1}, 2^{M-1}, and 2^{N+M-1} locations in the partial products. The outputs of the LUTs, O5 and O6, are provided to the carry chains as carry-generate and carry-propagate signals, respectively, to

Table 5.1 Comparing related works: Highlighted rows are also part of the contributions of the book

Article	Focus	Application-specific operators	Constraints for new operators/DSE			
			Accuracy	Resources	Latency	Power
[12]	Single multiplier + DSE	✗	✓	✓	✗	✓
[5]	Single multiplier + DSE	✗	✓	✓	✗	✗
[6]	Library of multipliers and adders	✗	✗	✗	✗	✗
[3]	Library of multipliers	✗	✗	✗	✗	✗
[10]	Library of multipliers + DSE	✓	✓	✓	✗	✗
[15]	DSE	✗	✓	✓	✓	✓
AppAxO	Library of multipliers + DSE	✓	✓	✓	✓	✓

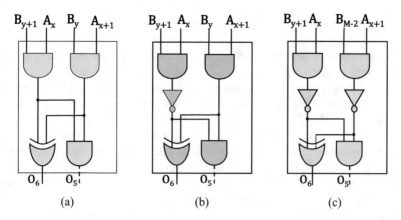

Fig. 5.3 Configurations of LUTs to implement an accurate multiplier. (**a**) Type-I. (**b**) Type-II. (**c**) Type-III

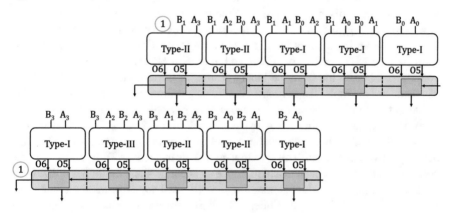

Fig. 5.4 LUT mapping of partial products for 4×4 accurate signed multiplier

generate intermediate results. These intermediate results, along with the two ①, are added together to compute the final accurate 4×4 product.

$$P = a_{N-1}b_{M-1}2^{N+M-2} + \sum_{n=0}^{N-2}\sum_{m=0}^{M-2} a_n b_m 2^{n+m} + 2^{N-1}\sum_{m=0}^{M-2} \overline{a_{N-1}b_m}2^m +$$

$$2^{M-1}\sum_{n=0}^{N-2} \overline{b_{M-1}a_n}2^n + 2^{N-1} + 2^{M-1} + 2^{N+M-1}$$

$$(5.1)$$

5.3.2   Approximation Methodology

The proposed approximation methodology is based on disabling LUTs (and the corresponding carry chain cells) in the accurate implementation to introduce approximations. For an M × N accurate multiplier, utilizing K LUTs for partial products generation, we have used a K-bit string (referred to as input configuration) to address each LUT. A $'0'$ at any location in the K-bit string represents the disabling of the corresponding LUT. Disabling a LUT means that the LUT will not contribute to producing the intermediate results. For this purpose, the carry-propagate signal (O6) for the corresponding carry chain cell is fixed to constant $'1'$. Since a disabled LUT will not contribute to intermediate results' value, the carry-generate signal is provided by the external bypass signal, and it is fixed to constant $'0'$. These settings enable the respective carry chain cell to forward the *preceding carry* without generating a new one. Further, the output of the corresponding carry chain cell (output of XOR gate) is truncated to $'0'$. These settings allow the synthesis tools not to use a disabled LUT during the synthesis and implementation process. For example, Fig. 5.5 shows the proposed approximation methodology for the 4 × 4 multiplier for a 10-bit configuration string $'1011011001'$. The LSBs $'11001'$ addresses the LUTs in the upper row, and the MSBs $'10110'$ identifies LUTs in the bottom row. For all the 0s in the binary string, the corresponding LUTs have been disabled. For example, for the least significant position in the second row, the O6 signal is fixed to $'1'$; therefore, the respective carry chain cell forwards the carry-in to the next carry chain cell. A bypass signal with the value $'0'$ is used instead of the O5 signal to help the synthesis tool not use the respective LUT. Further, the respective carry chain cell's output is also truncated to $'0'$ to show that the corresponding cell has been disabled. Since an accurate 4 × 4 multiplier utilizes *ten* LUTs for the partial product generation, the proposed approximation methodology supports 2^{10} approximate multipliers. Similarly, for an accurate 8 × 8 multiplier utilizing 36 LUTs for partial product generation, *AppAxO*'s proposed approximation

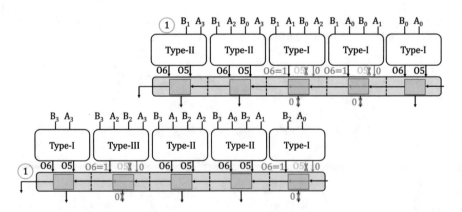

Fig. 5.5 4 × 4 approximate multiplier design according to input configuration "1011011001"

methodology provides 2^{36} different approximate multipliers with different accuracy and implementation performance metrics. It should be noted that the proposed approximation methodology is automated, generic, and scalable. Therefore, it can be utilized for implementing approximate circuits for any arithmetic operator (of arbitrary bit width) that utilizes LUTs and carry chains for its implementation.

5.3.3 Approximate Adders

The utilization of the carry chains in FPGA logic slices facilitates the implementation of an N-bit accurate adder using only N LUTs. Figure 5.6a represents the LUT mapping of an accurate unsigned 4-bit adder. For this purpose, the LUTs compute the required carry-generate (O5) and carry-propagate (O6) signals from the corresponding two input bits by performing logical *AND* and *XOR* operations on the input bits, respectively. The LUT configuration Type IV performs these operations in Fig. 5.6a. Further, to accommodate the overflows caused by adding two N-bit numbers, we produce an $(N + 1)$-bit output sum as shown in the figure. Utilizing the proposed *AppAxO* methodology, we use an N-bit string to address each LUT in the accurate implementation. A $'0'$ value at any position in the N-bit string (input configuration) denotes the disabling of the corresponding LUT and truncating the corresponding output generated by the respective carry chain cell to $'0'$. For instance, Fig. 5.6b represents an approximate 4-bit adder implementation for input configuration $'1101'$. As discussed previously, to disable the LUT at the

Fig. 5.6 LUT mapping of 4-bit adder. (**a**) Accurate implementation, (**b**) approximate adder design according to input configuration "1101"

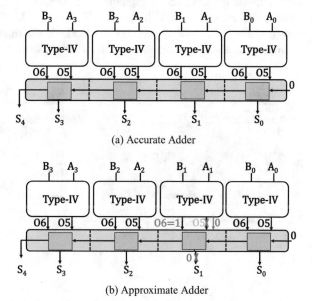

(a) Accurate Adder

(b) Approximate Adder

second least significant location, we utilize the external bypass signal to provide a $'0'$ as the carry-generate signal, and O6 provides a constant $'1'$ as the carry-propagate signal. For an N-bit accurate adder, *AppAxO* provides 2^N approximate adders with different accuracy and implementation performance metrics.

5.4 DSE for FPGA-Based Approximate Operators Synthesis

The proposed approximation methodology presents an opportunity for the designers to implement operators with improved Power, Performance and Area (PPA) if the application can tolerate the corresponding error profile of the operator. As described in the previous section, for an N-bit adder occupying only N LUTs, *AppAxO* provides 2^N different approximate adders. For example, for 4-bit, 8-bit, and 12-bit adders, *AppAxO* provides 2^4, 2^8, and 2^{12} approximate adder designs, respectively. Therefore, all approximate adders can be synthesized and implemented for an application due to adders' comparatively smaller design space. However, it can be noted that the proposed approximation methodology presents the designer with an exponentially increasing (with the bit width of multiplier) number of approximate multiplier choices to implement. For instance, the possible number of input configurations of the approximate multiplier increases from 1024 to nearly 64 billion as we consider 8-bit multipliers instead of 4-bits. In order to search for configurations that are appropriate for a given application, the designer can deploy a random search or generate configurations from intuition. However, such approaches do not provide a generalized methodology for designing application-specific approximate operators. Hence, to aid the designer in implementing the appropriate approximate multiplier design, we present two DSE methods—*AppAxO_MBO* and *AppAxO_ML*—as shown in Fig. 5.7. Both of the proposed methods use the automated modeling and estimation methodology described in the previous section.

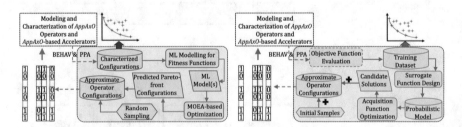

Fig. 5.7 Proposed design space exploration methods for *AppAxO*

5.4.1 DSE Using Bayesian Optimization

The left half of Fig. 5.7 shows the various stages involved in MBO-based approach to finding the set of Pareto front approximate multiplier—both for stand-alone multiplier designs and for any specific application. As described in Sect. 2.5, Bayesian optimization allows a more directed search that is particularly useful in case of problems involving costly fitness function evaluations. With *AppAxO_MBO*, we present a multi-objective optimization method that uses the input configurations of the approximate multiplier as the design variables for a bi-objective optimization problem. We use separate probabilistic models for each design objective—one that quantifies the application/multipliers behavioral accuracy and the second quantifying the PPA design intent. The implementation of the acquisition function involved generating a set of random input configurations, using the surrogate function to predict their fitness, then ranking them according to their expected contribution to the Pareto front hypervolume, and selecting a fixed number of top-ranked samples for the next iteration.

5.4.2 MOEA-Based Optimization

In addition to AppAxO_MBO, we present a Multi-objective Evolutionary Algorithm (MOEA)-based approach to the DSE problem of *AppAxO*. Specifically, we use GA for the bi-objective optimization problem. As discussed in Sect. 2.5, MOEA-based optimization methods usually rely on low-cost fitness function evaluation, and hence, a naive approach to implementing GA would limit the usability in *AppAxO* to small designs with low synthesis time-cost. To this end, as shown in the right half of Fig. 5.7, we present AppAxO_ML, an ML-based DSE approach. We use an initial set of random samples to train multiple ML models for predicting PPA and accuracy metrics of an application. The ML-based models are in turn used during GA-based search to predict the Pareto front design points, the Predicted Pareto Front (PPF). The collection of PPFs from each ML model are then evaluated and filtered to provide the Evaluated Pareto Front (EPF). The details of the ML models used in AppAxO_ML are described in the next section.

5.4.3 Machine Learning Models for DSE

The true evaluation of the application's behavioral accuracy and the accelerator's PPA metrics can be a bottleneck in the randomized algorithm-based DSE methods. For instance, the synthesis of the accelerator for 2D convolution of an image of size 128×128 can take up to 20 min on a standard computer. To this end, ML-based prediction is used in the fitness evaluation of AppAxO_ML. We use the configuration vector that defines the position of the LUTs used in the approximate multiplier as the input to the ML models. We use a separate model for each behavioral accuracy and PPA metric, and the ML models are then used to predict the accuracy, power, CPD, and LUT utilization for any arbitrary configuration. The following models were used with their most widely used configurations.

- Random Forest Regression (RFR): It is an ensemble learning-based supervised algorithm. The ensemble technique utilizes multiple ML models for predictions. These predictions are then combined to make more accurate predictions than a single ML model. It operates by constructing a multitude of decision trees at training time [17]. Each tree draws a random sample from the original dataset when generating its splits, adding an additional randomness element that prevents overfitting of the model.
- Support Vector Regression (SVR): Compared to a simple linear regression, SVR allows the specification of an acceptable error margin and tolerance of a model for the specified error margin [18]. However, SVR requires specific calibrations of different features, which makes it less robust.
- Stochastic Gradient Descent (SGD): It is an optimization technique and is generally used in sparse ML problems. The gradient of the loss is estimated (each sample at a time), and the model is updated accordingly with a decreasing learning rate [19]. It is, however, susceptible to feature scaling and requires tuning of several hyperparameters.
- Gradient Boosting Regression (GBR): In this technique, a predictive model is produced from an ensemble of weak predictive models [20]. It is similar to RFR; however, the individual predictive models are combined at the start to produce an output, whereas it is done at the end for RFR.
- Decision Tree Regression (DTR): These are a series of sequential steps designed to solve a problem and provide probabilities, costs, or other consequences of making a particular decision. Smaller subsets are broken down from a dataset, while the associated decision tree is incrementally developed at the same time [21].
- Multi-layer Perceptron (MLP): It is a class of feedforward ANNs and uses backpropagation for training. MLP learns a function between the input features and the output by using intermediate hidden layers [22].

5.5 Results and Discussion

5.5.1 Experimental Setup and Tool Flow

We have implemented all presented multipliers in VHDL and synthesized them for the $7VX330T$ device of the Virtex-7 FPGA family using Xilinx Vivado 19.2. As discussed in the Experimental Setup of previous chapters, we have implemented each operator—adder and multiplier—design multiple times to obtain precise critical path delay and dynamic power consumption values. Using this characterization method, the accuracy and PPA estimation for each approximate 8×8 multiplier configuration consumes nearly 3.55 min of processing time. The accelerators for the applications were implemented using different high-level languages. For the calculation of the dynamic power of all implementations, Vivado Simulator and Power Analyzer tools have been utilized. All applications have been implemented for Xilinx Zynq UltraScale+ MPSoC (xczu3eg-sbva484-1-e device). All behavioral estimations were implemented in Python. The ML-based modeling, and the MBO-based DSE methodology were implemented in Python using multiple packages, including scikit-learn, TensorFlow [23], and PyGMO [24]. All experiments were conducted on an HPC server with a single AMD EPIC^{TM} processor with 24 cores and two PCIe Gen4 NVIDIA A100 Tensor-Core-GPUs, with 512GB of DDR4-3200MHz main memory. It must be noted that we have used homogeneous accelerator designs for each application. Therefore, all the approximate multiplier designs used in any arbitrary accelerator use similar LUT configurations. The following applications were used in the experiments for demonstrating the effectiveness of *AppAxO*.

5.5.1.1 ECG Peak Detection

To represent the set of applications using one-dimensional convolution, we use the peak detection of ECG signal using Pan-Tompkins algorithm [25] as the test application. Pan-Tompkins algorithm consists of five stages—*low-pass filtering, high-pass filtering, derivative filtering, squaring, moving window*, and *peak finder*. For the current work, we explored the effect of approximate multipliers in the first stage—the low-pass filter with a cutoff frequency of 15Hz and having 11 *taps*. Both the signals and the filter coefficients were quantized to integer values corresponding to the range of the multiplier's operands. Although the accuracy of the implementation is characterized by multiple interpretations of the confusion matrix, we use only *accuracy* as the relevant metric for our current work. The accuracy is evaluated as the ratio of the *true positives* and the sum of *true positives, false negatives*, and *false positives* of the detected peaks. The behavioral results are reported for 100 signals from the PhysioNet's single-lead ECG dataset [26]. For the accelerator implementation of the low-pass Finite Impulse Response (FIR) filter, we used Chisel3 [27]. The FIR low-pass filter designed takes 11 points in the signal data as normalized integer data and approximately multiplies it with the coefficients of the 11-window integer-normalized coefficients. The resulting accumulation of the products is quantized back into floating-point values for further processing. The behavioral estimation for a single test case of ECG peak detection using 8×8 approximate multiplier configuration expends an average of 1.5 min of processing time. Similarly, the accelerator synthesis and implementation consume 1.73 min of processing time.

5.5.1.2 Gaussian Smoothing

We use a Gaussian Smoothing (GS), a frequently used benchmark for representing the set of 2D convolution-based applications. It involves a moving window of GS filter coefficients and is generally used to remove high-frequency noise from images. In *AppAxO*, we implemented a 5×5 kernel, and the accelerator, implemented using Vivado HLS, uses a line buffer, along with 25 multipliers. Similar to ECG, the kernel coefficients and inputs were quantized to the multipliers' bit width. We use the average reduction in the PSNR as the behavioral minimization objective. It represents the negative of the PSNR improvement using GS with an approximate multiplier over 45 images. The PSNR metrics for both noisy and processed images were computed compared to a baseline noise-free image. The behavioral estimation for a single test case of GS using 8×8 approximate multiplier configuration expends an average of 3.35 min of processing time. Similarly, the accelerator synthesis and implementation consume 5.06 min of processing time.

5.5.1.3 MNIST Digit Recognition

Image classification is a commonly used application for evaluating the efficacy of various approximation techniques. We have used a lightweight MLP implemented in Python for the classification of the MNIST Digit dataset [28]. The MLP consists of two fully connected hidden layers having 100 and 32 nodes, respectively. The input and output layers have 784 and 10 nodes, respectively, to classify the 28×28 images of the dataset into *ten* different classes. The network is trained with 50,000 training images using IEEE double-precision numbers. For this work, we have evaluated the efficacy of the various approximate multipliers by deploying them in the output layer of the network for inference using 10,000 test images. For this purpose, we have quantized the trained weights and input activations of the last layer according to the multiplier size under consideration. To compute the application's performance metrics, we have implemented the last layer in Vivado HLS and evaluated for different approximate multipliers. We shall refer to this application as MNIST in the discussion of the experiment results. The behavioral estimation for a single test case of MNIST using 8×8 approximate multiplier configuration expends an average of 0.78 min of processing time. Similarly, the accelerator synthesis and implementation consume 2.52 min of processing time.

The approximation-aware DSE for both operator-level and application-level design involves finding multiple design points that provide varying levels of behavioral (accuracy) and PPA trade-offs. For our current work, we use application-specific accuracy metric and the product of PDP and LUTs utilization as the two objectives. Consequently, the DSE runs for the experiments involve multi-objective optimization (with two objectives). Therefore, we use hypervolume of the Pareto front and the number of non-dominated design points, two commonly used metrics for comparing the results from different multi-objective optimization runs. As described in Chap. 2, for a two-objective problem, the hypervolume corresponds to the area between the non-dominated Pareto front and a *reference point*. For our current work, we aim at minimizing both the approximation-induced error metric and PDP $\times$ LUT. Hence, the reference point comprises of the maximum of both the metrics across all Pareto front points under consideration.

5.5.2 Accuracy-Performance Analysis of Approximate Adders

For an adder utilizing N LUTs and corresponding carry chain elements, our proposed approximation methodology generates 2^N approximate adder designs. The adder having input configuration $2^N - 1$ utilizes all LUTs, and it is an accurate adder. For example, for a 4–bit and 12–bit adder, our methodology generates 2^4 and 2^{12} different approximate adders, respectively. For example, Fig. 5.8 compares the hardware performance metrics and accuracy of the *AppAxO-* generated 15 4-bit approximate adders. Please note that we do not synthesize hardware for configuration $'0000'$ (all LUTs disabled). For assessing the performance of the designs, we have used the PDP $\times$ LUT metric to incorporate all design metrics. A smaller value of PDP $\times$ LUT represents a circuit with better performance. Similarly, we have used three different error metrics for accuracy evaluation of the individual adders, i.e., average absolute error, average absolute relative error, and error probability. Figure 5.8a shows that a total of *five* design configurations ($'1111', '1110', '1101', '1100',$ and $'1000'$) lie on the Pareto front. The non-dominated adder configuration $'1111'$ utilizes all four LUTs and is an accurate adder. Moreover, all non-dominated design points have the most significant LUT enabled. An interesting design point is the adder configuration $'1000'$, which utilizes only a single LUT. Figure 5.8b compares the average absolute relative error and PDP $\times$ LUT of the 15 design points. The analysis returns the same five non-dominated designs as described in Fig. 5.8a. The analysis in Fig. 5.8c shows a total of 12 non-dominated design points. These points also include four non-dominated design points from Fig. 5.8a.

5.5.3 Accuracy-Performance Analysis of Approximate Multipliers

For a multiplier utilizing K LUTs for partial product generation, our proposed approximation methodology generates 2^K different approximate multipliers. The multiplier having input configuration $2^K - 1$ utilizes all LUTs, and it is an accurate multiplier . For example, for a 4×4 and an 8×8 multiplier, our methodology generates 2^{10} and 2^{36} different approximate multipliers, respectively.

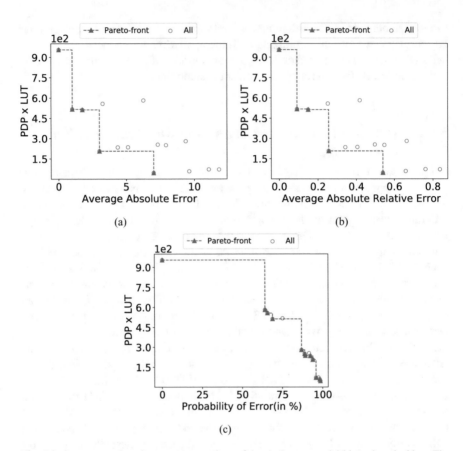

Fig. 5.8 Accuracy and performance comparison of *AppAxO*-generated 4-bit unsigned adders. The power and CPD are in μW and ns, respectively

5.5.3.1 Multiplier-Level Analysis

To evaluate the efficacy of the generated multipliers, Fig. 5.9 presents an accuracy-performance analysis of all the 4×4 generated multipliers. For the analysis, we have excluded multiplier configuration $'0000000000'$, which disables all the LUTs in the partial product generation stage.[1] For evaluating the performance of the multipliers, we have used the PDP $\times$ LUT metric.

Figure 5.9a analyzes the average absolute error and PDP $\times$ LUT metric of all 1023 multipliers. The analysis shows that a total of 38 multiplier designs lie on the Pareto front. The non-dominated multiplier configuration $'1023'$ (binary value $'1111111111'$) utilizes all ten LUTs for partial products generation and produces 0 average absolute error. Most of the non-dominated design points have LUTs enabled

---

[1] The mapping of multiplier configuration (a binary string) to LUTs is described in Fig. 5.5.

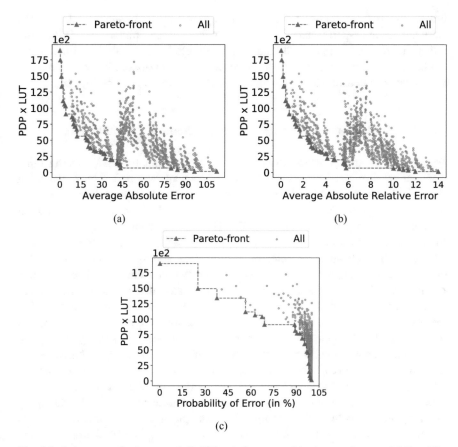

Fig. 5.9 Accuracy-performance analysis of *AppAxO*-generated 4×4 approximate multipliers. The power and CPD are in μW and ns, respectively

at most significant locations. An interesting observation is the non-dominated multiplier designs that enable only a single LUT for partial product generation. For example, non-dominated multiplier designs with configurations 2, 4, 8, 1, and 512 enable only a single LUT for their partial products. Designs 1 and 512 have the binary configuration $'0000000001'$ and $'1000000000'$, respectively. Design 1 utilizes only the least significant LUT in the first partial product row, and design 512 deploys only the most significant LUT in the second row of partial products. Design 512 has a lower average absolute error and higher PDP × LUT than design 1.

Figure 5.9b presents the Pareto analysis of the 1023 design points for PDP × LUT and average absolute relative error metrics. Compared to the non-dominated designs in Fig. 5.9a, average absolute relative error introduces five new multipliers to the set of Pareto designs. Similarly, the comparison in Fig. 5.9c shows a total of 20 non-dominated multipliers out of 1023 designs. These points also include 11 new multiplier designs, which were dominated in the other two plots. However, the generated approximate multipliers' application-specific efficacy cannot be determined from the accuracy-performance analysis of individual multipliers. For this purpose, either the generated approximate multipliers should be exhaustively utilized in an application's implementation, or some machine learning-based intelligent models should be used to estimate the potential efficiency of the various multipliers.

5.5.3.2 Application-Level Analysis of Approximate 4 × 4 Multipliers

Figure 5.10 shows the utilization of all 4 × 4 approximate multipliers[2] in three different applications. The output accuracy of each application has been computed by utilizing the multipliers' behavioral models in the high-level implementation of each application. Similarly, the performance metric (PDP × LUT) has been computed by utilizing the multipliers' VHDL implementation in the accelerator of each application. Figure 5.10a shows the accuracy-performance analysis of the ECG application. A total of 12 different approximate multiplier-based designs are non-dominated design points with different accuracy and performance parameters. Six designs among these non-dominated designs utilize multipliers which were among dominated points in Fig. 5.9. It is interesting to note that the accurate multiplier-based accelerator is not among the non-dominated design points. The ECG application's inherent error tolerance significantly masks the approximations-generated errors and removes the accurate multiplier-based design from the set of Pareto points. The approximate multiplier with binary configuration '1111101111' (decimal value 1007) produces the highest output accuracy. Moreover, the application-specific accuracy-performance analysis can reveal highly efficient multipliers for that application. For example, approximate multipliers with configuration values 16, 128, and 512 utilize a single LUT for partial product generation; however, their corresponding accelerators are among non-dominated design points. Figure 5.10b and c show the utilization of the approximate multipliers for the Gaussian smoothing filter and MNIST dataset classification MLP, respectively. For both applications, a total of 13 different approximate multiplier-based designs lie on the Pareto fronts. Further, for both applications, the accurate multiplier-based designs are not among the non-dominated accelerator designs. Similar to the ECG application's error resilience, the Gaussian smoothing filter and the MNIST classification expose multipliers that utilize only a single LUT for partial product generation. For example, for the Gaussian smoothing filter, approximate multipliers with configuration values 1, 4, and 16 generate non-dominated design points. These application-specific accuracy-performance analyses augment the need for designing approximate multipliers according to application accuracy-performance requirements. We have also evaluated the efficacy of the *AppAxO* methodology for 8 × 8 multipliers using various ML models. These results are described in the following subsections.

---

[2] Excluding multiplier configuration with binary value '0000000000'.

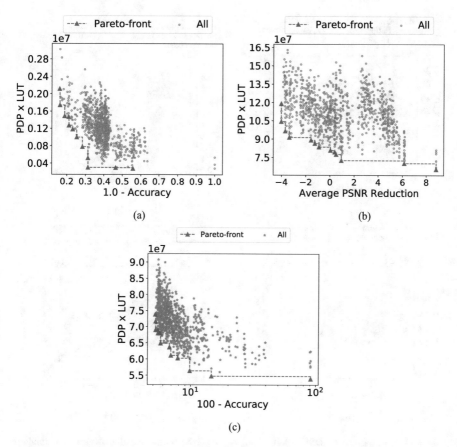

Fig. 5.10 Accuracy-performance analysis of approximate 4 × 4 multipliers in three different applications. The power and CPD are in μW and ns, respectively. (**a**) ECG. (**b**) GS. (**c**) MNIST

5.5.4 AppAxO_MBO

To show the effectiveness of the MBO-based exploration, we compared the Pareto front obtained by AppAxO_MBO starting with 100 random samples to that obtained by the initial random samples in the search for 8 × 8 approximate multipliers. The DSE run configuration includes 200 iterations with 1000 acquisition samples and 10 true evaluations per iteration. It must be noted that the true evaluations include actual synthesis and implementation of the multipliers' accelerators along with Python-based behavioral estimation for each configuration. As a result, the processing time for each DSE run consumed nearly a week of processing time. Figure 5.11 shows the results in the search for stand-alone multipliers and that for the three test applications. The bar plots in each subfigure show the comparison of the hypervolume, and the labels on top of each bar show the number of points

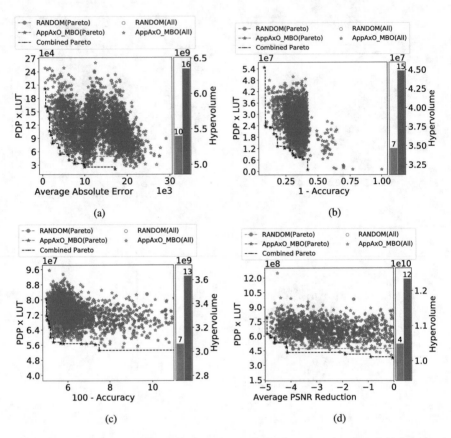

Fig. 5.11 Comparison of DSE results from AppAxO_MBO against that from initial random samples. (**a**) MULT 8×8. (**b**) ECG. (**c**) MNIST. (**d**) GS

in the corresponding Pareto fronts. As seen on the figure, the MBO-based DSE resulted in improved hypervolume for all cases, with maximum benefits observed for ECG. However, the MBO-based search results in a large number of points being synthesized, many of which are not on the Pareto front.

5.5.5 ML Modeling

Our proposed approximation methodology generates 2^{36} approximate multipliers for 8×8 multiplication. To explore the large design space of the 8×8 multipliers and their utilization in the three test applications, we have used the ML regression algorithms described in Sect. 5.4.3. We have trained separate models for multipliers and each application to estimate the corresponding accuracy and performance metrics. To evaluate the efficacy of our models, we have used the fidelity metric. The fidelity metric denotes the relationship $(=, <, >)$ between the input configurations and their corresponding actual and predicted output values. The lower the Fidelity Error (FE), the higher the correspondence between the predicted metric value and the actual value. To further assess our models' efficiency, we also report the Mean Square Error (MSE) and Mean Absolute Error (MAE)—average absolute error—of the models' predictions.

For training and testing our models, we have used four datasets[3] of 2060 configurations each. Each configuration shows the impact of a unique approximate multiplier[4] on the output accuracy and the corresponding performance metrics. Our model employs randomly chosen 80% configurations of the input dataset for training the model. The remaining 20% of the input dataset is used for testing the trained model. The results of ML models giving the minimum error for each of the metrics across different applications have been tabulated in Table 5.2. The processing time needed for the training and the average inference time for each configuration is shown in Tables 5.3 and 5.4. The inference time is averaged over the total time consumed in performing inference for the complete dataset. We see the performance of these ML models on the following metrics:

- 8×8 multiplier performance metrics (Fig. 5.12)—We see that test FE for average absolute relative error metric is lowest for MLP at 0.91%, followed by GBR and RFR. The test MSE and MAE are also lowest for MLP regressor, with each being 0.52 and 0.50, respectively. For CPD, we see the test-FE scores are lowest for SGD at 16.41%, with test-MSE of 0.017 and test-MAE of 0.10 being the lowest for SVR. For power and LUTs estimations, MLP gives a good performance with test-FE of 2.95%, test MSE of 753.81, and test-MAE of 21.99 for power values. SGD offers the best test-FE of 2.62%, and SVR provides a good test-MSE of 0.81 and test-MAE of 0.71 for LUTs.
- ECG Performance Metrics—We see that test FE of Accuracy is the lowest for RFR at 3.86%. The test-MSE and test-MAE are also lowest for RFR, with each of them being 0.00041 and 0.01, respectively. For CPD, we see the test-FE is lowest for DTR at 23.23%, test-MSE lowest for RFR of 0.02, and test-MAE for RFR of 0.13. GBR gives low test-FE for Power at 4.56%. The lowest test-MSE of 1585410.54, and the lowest test-MAE of 922.17, are observed for GBR. For LUTs, a good performance is observed with MLP with test-FE of 2.85%, test-MSE as 105.93, and 8.21 as test-MAE.
- MNIST performance metrics—We see that test FE of accuracy is the lowest for RFR at 6.30%. The test-MSE is lowest for MLP at 2.53, and test-MAE with RFR at 0.71 is the lowest. For CPD, we see the test-FE is lowest for GBR at 20.31%, and test-MSE and test-MAE are lowest for RFR at 0.45 and 0.52, respectively. RFR gives low test-FE for power at 17.75%. The lowest test-MSE of 12.30 is observed for GBR, and the lowest test-MAE of 2.23 is observed for DTR. For

---

[3] One for multipliers and one for each application.

[4] 8×8 multipliers use a 36-bit string to represent LUTs for partial products generation as described in Sect. 5.3.2.

Table 5.2 The ML models reporting the minimum error in terms of Mean Square Error (MSE), Mean Absolute Error (MAE), and Fidelity Error (FE) (in %) for the three applications. ML models used: Random Forest Regression (RFR), Support Vector Regression (SVR), Stochastic Gradient Descent (SGD), Gradient Boosting Regression (GBR), Decision Tree Regression (DTR), and Multi-layer Perceptron (MLP)

Metric	Minimum MSE (Train, test)			Minimum MAE (Train, test)			Minimum FE (Train, test)		
Application	ECG	MNIST	GS	ECG	MNIST	GS	ECG	MNIST	GS
App-specific accuracy	RFR, RFR	RFR, MLP	RFR, MLP	RFR, RFR	RFR, RFR	RFR, GBR	RFR, RFR	RFR, RFR	RFR, GBR
CPD [nS]	RFR, RFR	RFR, RFR	RFR, RFR	RFR, RFR	RFR, RFR	RFR, GBR	RFR, DTR	RFR, GBR	RFR, SGD
Power [μW]	RFR, GBR	RFR, GBR	SGD, SGD	RFR, GBR	RFR, DTR	SGD, SGD	RFR, GBR	SVR, RFR	RFR, SGD
LUT utilization	SGD, MLP	SGD, SGD	SGD, SGD	MLP, MLP	SGD, SGD	SGD, SGD	RFR, MLP	RFR, SGD	RFR, SGD

Table 5.3 The execution time for training and inference on various ML models for different target metrics. All the timing values are reported in *milliseconds*. The inference timing values are reported for a single data point averaged over the inference of the whole training set (2060 points). ML models used: Random Forest Regression (RFR), Support Vector Regression (SVR), Stochastic Gradient Descent (SGD), Gradient Boosting Regression (GBR), and Multi-layer Perceptron (MLP)

Application	Target metric for ML model	GBR		MLP		RFR	
		Train	Infer	Train	Infer	Train	Infer
ECG	Accuracy	2023.721	0.377	242.924	0.175	698.487	6.05
	CPD	2060.872	0.403	708.342	0.244	751.05	7.214
	Power	2131.917	1.215	4505.54	0.179	1010.047	8.782
	LUTs	2154.619	0.403	3673.557	0.179	678.572	6.056
GS	Accuracy	2004.419	0.334	2857.172	0.177	706.297	5.789
	CPD	2331.065	0.422	2614.81	0.186	706.558	6.391
	Power	2259.041	0.463	4592.353	0.197	712.243	6.289
	LUTs	2130.301	0.469	5095.321	0.198	708.136	5.961
MNIST	Accuracy	2033.754	0.42	4296.751	0.171	692.762	5.878
	CPD	2127.168	0.588	2581.044	0.197	702.273	6.042
	Power	2397.082	0.51	5140.923	0.186	925.009	7.447
	LUTs	1993.787	0.342	4202.479	0.174	762.154	5.865
MULT 8 × 8	Average absolute error	2158.514	0.329	4462.492	0.182	792.243	6.247
	Average absolute relative error	2014.313	0.343	3820.165	0.169	689.845	6.253
	CPD	2883.947	0.393	889.988	0.176	769.147	8.08
	Power	2038.58	0.456	3105.816	0.178	693.654	5.665
	LUTs	2185.196	0.51	4232.282	0.207	685.203	6.134

LUTs, a good performance is observed with SGD with test-FE of 2.89%, test-MSE as 1415.29, and 29.9 as test-MAE.

- GS performance metrics—We see that test FE of accuracy is the lowest for GBR at 2.94%. The test-MSE is lowest for MLP at 0.25, and test-MAE is lowest for GBR at 0.34. For CPD, we see the test-FE is lowest for SGD at 17.40% with test-MSE of 0.04 with RFR, and test-MAE is lowest for GBR at 0.15. SGD gives low test-FE for power at 3.17%. The lowest test-MSE of 83177.84 and the lowest test-MAE of 230.61 are observed for SGD. For LUTs, the best performance is observed with SGD with test-FE of 2.89% test-MSE as 601.65 and 19.70 as test-MAE.

Based on the above discussion, it can be concluded that different ML models provide varying levels of accuracy in predicting the performance metrics of an application. Further, as shown in Tables 5.3 and 5.4, the inference time for each model can vary across a wide range. The designer can make a decision regarding the choice of ML model to use for DSE based on these two factors—accuracy and processing (inference) time. As we shall show next, collecting the results from DSE runs with varying types of ML models provides the best Pareto front for

Table 5.4 The execution time for training and inference on various ML models for different target metrics. All the timing values are reported in *milliseconds*. The inference timing values are reported for a single data point averaged over the inference of the whole training set (2060 points). ML models used: Random Forest Regression (RFR), Support Vector Regression (SVR), Stochastic Gradient Descent (SGD), Gradient Boosting Regression (GBR), and Multi-layer Perceptron (MLP)

Application	Target metric for ML model	SGD		SVR	
		Train	Infer	Train	Infer
ECG	Accuracy	4.185	0.097	14.323	0.103
	CPD	6.456	0.097	225.132	0.179
	Power	24.82	0.096	241.39	0.219
	LUTs	34.46	0.105	245.992	0.237
GS	Accuracy	12.686	0.096	262.942	0.186
	CPD	33.315	0.128	221.338	0.19
	Power	55.562	0.097	250.689	0.248
	LUTs	40.898	0.105	237.082	0.219
MNIST	Accuracy	36.383	0.093	264.139	0.187
	CPD	29.538	0.099	274.11	0.203
	Power	116.668	0.121	272.108	0.22
	LUTs	46.64	0.094	243.259	0.196
MULT 8 × 8	Average absolute error	34.367	0.103	235.551	0.201
	Average absolute relative error	31.939	0.098	258.572	0.215
	CPD	28.563	0.215	205.164	0.25
	Power	9.651	0.1	232.453	0.2
	LUTs	10.355	0.097	262.186	0.197

each application. However, in a time-constrained scenario, using a selection of ML models (based on their fidelity error performance) can provide results that are close to the results collected from multiple DSE runs using varying models.

5.5.6   DSE Using ML Models

The ML models described in the earlier section were used in the GA-based DSE for 8 × 8 approximate multipliers. The experiments included six sets of DSE runs. Five experiments involved using each of the ML across every performance metric prediction. The sixth experiment, *SELECT*, involved choosing the model with the lowest fidelity test error for each performance metric. DTR experiments are not reported since all evaluated points using DTR were reported as Pareto front points. Tables 5.5 and 5.6 show the processing time for executing the GA-based DSE using the ML models for fitness evaluation across all the experiments. The GA-specific configurations for the DSE experiment include 200 generations with starting population of 1000 samples; mutation and cross-over probability of 0.04 and 0.8, respectively; and tournament-based selection with a tournament size of 3.

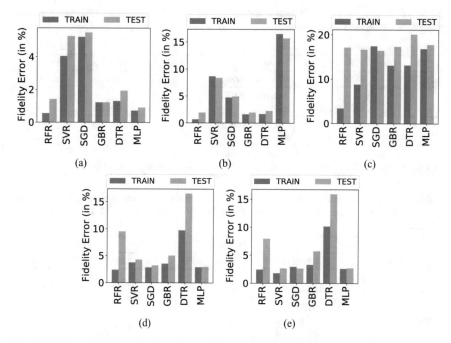

Fig. 5.12 FE in the ML models for different performance metrics of 8 × 8 multipliers. (**a**) Average absolute relative error, (**b**) average absolute error, (**c**) CPD (in nS), (**d**) power (in μW), (**e**) LUT utilization

Table 5.5 Processing time (in minutes and seconds) for running GA-based DSE using the ML models for the fitness estimation

Application	PPA	Behavioral	GBR	MLP	RFR
ECG	PDP × LUT	1—Accuracy	3m18s	1m27s	55m12s
MNIST	PDP × LUT	100—Accuracy	3m47s	1m29s	54m23s
GS	PDP × LUT	Average PSNR reduction	3m22s	1m27s	54m13s
MULT 8 × 8	PDP × LUT	Average absolute error	6m34s	2m19s	94m34s
MULT 8 × 8	PDP × LUT	Average absolute relative error	5m51s	2m19s	96m6s

Each of the DSE experiments resulted in a Predicted Pareto Front (PPF), where the metrics correspond to that obtained using ML-based predictions. Then, the set of approximate multiplier configurations from each PPF were evaluated with actual hardware synthesis and behavioral testing, followed by Pareto front determination with the actual metrics to obtain the Evaluated Pareto Front (EPF). Tables 5.7 and 5.8 show the number of points in PPF and EPF of each experiment. As expected, not all points in PPF translated to actual Pareto front design points in the EPF. Figure 5.13 shows the EPF (evaluated) and the PPF (predicted) for the experiments

Table 5.6 Processing time (in minutes and seconds) for running GA-based DSE using the ML models for the fitness estimation

Application	PPA	Behavioral	*SELECT*	SGD	SVR
ECG	PDP × LUT	1—Accuracy	14m51s	0m43s	1m31s
MNIST	PDP × LUT	100—Accuracy	27m16s	0m43s	1m42s
GS	PDP × LUT	Average PSNR reduction	1m24s	0m43s	1m42s
MULT 8 × 8	PDP × LUT	Average absolute error	15m48s	0m56s	2m39s
MULT 8 × 8	PDP × LUT	Average absolute relative error	15m57s	0m58s	2m41s

Table 5.7 DSE-level evaluation of ML models. The number of design points in the Predicted Pareto Front (PPF) and Evaluated Pareto Front (EPF) is reported for the DSE runs for six experiments

Application	Objectives		GBR		MLP		RFR	
	PPA	Behavioral	PPF	EPF	PPF	EPF	PPF	EPF
ECG	PDP × LUT	1—Accuracy	30	11	34	10	19	10
MNIST	PDP × LUT	100—Accuracy	39	13	15	9	31	11
GS	PDP × LUT	Average PSNR reduction	32	10	34	13	39	13
MULT 8 × 8	PDP × LUT	Average absolute relative error	25	17	29	20	35	22
MULT 8 × 8	PDP × LUT	Average absolute error	28	21	7	4	46	28

with ECG for all six sets of experiments. RFR and *SELECT* show the closest match between the PPF and the EPF.

To evaluate the quality of results with each model, we show the EPFs along with a combination of the EPF (All) for the experiments in Fig. 5.14. The numbers in the parenthesis of each label correspond to the contribution of each model to the combined Pareto front. The hypervolume of all the models, along with the combined results (ALL), is shown in Fig. 5.15 for different application and multiplier DSE experiments. It can be observed that ideally, the combination of results from each of the ML models along with the selective model assignment (*SELECT*) results in the highest quality of results.

Figure 5.16 shows the comparison of the Pareto front of the designs obtained from three different ways. *AppAxO_RND* denotes the set of 2048 randomly generated design configurations for the approximate multiplier. Additionally, 12 configurations denoting corner-case designs were added to *AppAxO_RND* to

Table 5.8 DSE-level evaluation of ML models. The number of design points in the PPF and EPF is reported for the DSE runs for six experiments

Application	Objectives PPA	Behavioraal	SGD PPF	EPF	SVR PPF	EPF	*SELECT* PPF	EPF
ECG	PDP × LUT	1—Accuracy	87	15	37	13	27	10
MNIST	PDP × LUT	100—Accuracy	43	10	28	8	17	8
GS	PDP × LUT	Average PSNR Reduction	14	10	49	9	17	11
MULT 8 × 8	PDP × LUT	Average absolute relative error	22	13	32	16	27	21
MULT 8 × 8	PDP × LUT	Average absolute error	30	20	30	15	27	20

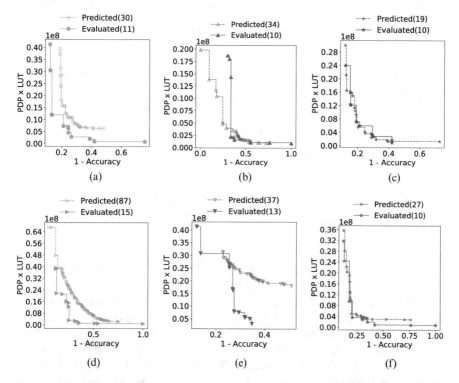

Fig. 5.13 Comparison of the accuracy of the prediction of Pareto front design points predicted by the ML models for ECG. Models: (**a**) Gradient Boosting Regression (GBR), (**b**) Multi-layer Perceptron (MLP), (**c**) Random Forest Regression (RFR), (**d**) Stochastic Gradient Descent (SGD), (**e**) Support Vector Regression (SVR), (**f**) *SELECT*: uses the model with the lowest testing Fidelity Error (FE) for each metric

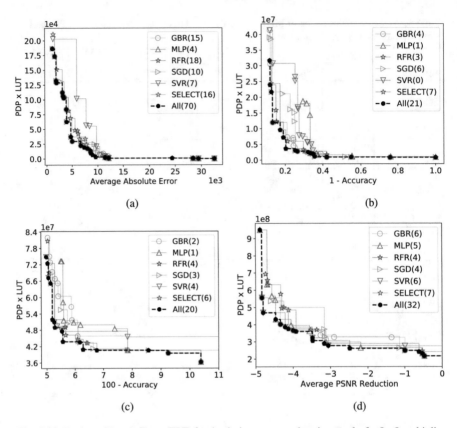

Fig. 5.14 Evaluated Pareto Front (EPF) for the design space exploration results for 8 × 8 multiplier and different applications using ML models. "All" refers to the combined Pareto front. (**a**) MULT 8 × 8. (**b**) ECG. (**c**) MNIST. (**d**) GS

generate the set of 2060 design points, *AppAxO_TRN*, that were used for the training of the ML models. One of the corner-case designs is the configuration ′1111111111111111111111111111111111111111′, where all LUTs are enabled for the multiplier. The remaining 11 corner-case configurations were selected intuitively from the insight of the application-level analysis of 4 × 4 multipliers. For example, the application-level analysis of 4 × 4 multipliers had a few non-dominated design points where only a single LUT was enabled for the deployed multiplier. Therefore, for *AppAxO_TRN*, we had selected such 8 × 8 multiplier configurations where only one, two, or three LUTs were enabled (either at the most significant or least significant locations). For example, the 8 × 8 multiplier configurations ′1000000000000000000000000000000000000000′ and ′0000000000000000000000000000000000000001′ are two such configurations with one LUT enabled at the most significant and least significant locations, respectively.

Finally, *AppAxO_ML* denotes the set of design points generated using the proposed ML-based DSE. The design points for AppAxO_ML include the collection

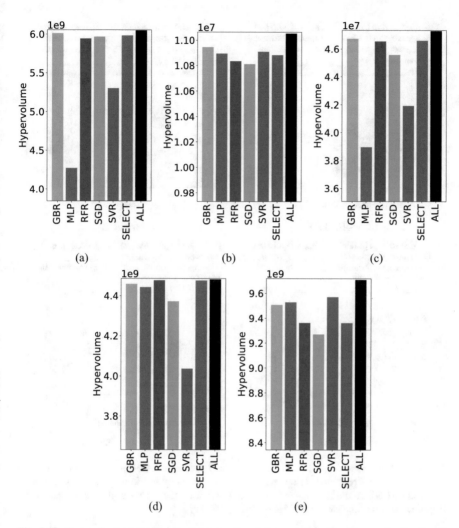

Fig. 5.15 Comparison of the hypervolume for the design space exploration results for 8×8 multiplier and different applications using ML models. "All" refers to the hypervolume of the combined Pareto front. (**a**) MULT 8×8 with average absolute error as the accuracy metric, (**b**) MULT 8×8 with average absolute relative error as the accuracy metric, (**c**) ECG, (**d**) MNIST, (**e**) GS

of predicted Pareto front points from the DSE runs corresponding to GBR, MLP, RFR, SGD, SVR, and *SELECT*. It can be observed that in all cases, the ML-based DSE results in higher hypervolume and more number of Pareto front points. The combined Pareto fronts included 9, 4, 3, and 2 design points with AppAxO_RND; 14, 6, 7, and 3 design points with AppAxO_TRN; and 25, 16, 13, and 19 design points with AppAxO_ML for MULT 8×8, ECG, MNIST, and GS, respectively.

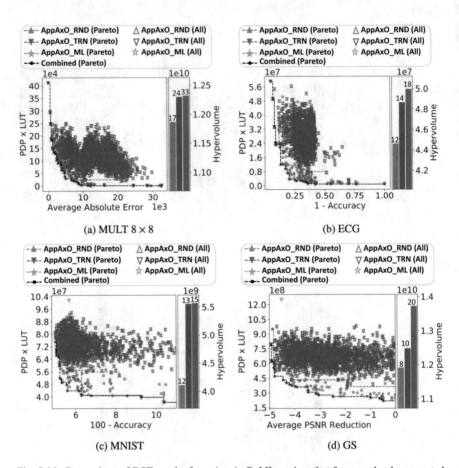

Fig. 5.16 Comparison of DSE results from AppAxO_ML against that from randomly generated points (AppAxO_RND) and design points used for training and testing of the ML models (AppAxO_TRN). Additional points synthesized, based on the Pareto front points reported by AppAxO_ML, for each case: (**a**) 115, (**b**) 216, (**c**) 154, (**d**) 146

It can be observed that in some cases (MULT and MNIST), AppAxO_ML did not result in significant improvements over AppAxO_TRN. In both these cases, the intuitively added corner-case design points sufficed in providing the requisite accuracy-performance trade-offs. However, in other cases, AppAxO_ML succeeded in providing considerably better Pareto front design points. Therefore, AppAxO_ML can aid the designer in searching for design points, beyond the more generic corner-case designs, that exploit the application's inherent error tolerance. Moreover, unlike in the case of AppAxO_MBO (Fig. 5.11), the candidate solutions generated by AppAxO_ML for true characterization are closer to the final Pareto front.

5.5.7 Proposed Approximate Operators

5.5.7.1 Approximate Adders

To evaluate the efficacy of the *AppAxO*'s modeling of approach for generating approximate arithmetic operators, Fig. 5.17 compares the hardware performance metrics and accuracy of *AppAxO*-generated 12-bit approximate unsigned adders with the 32 unsigned designs of the ApproxFPGAs library [16]. For a fair comparison, we have resynthesized and re-implemented all the 12-bit approximate adders of ApproxFPGAs. For a 12-bit accurate adder, *AppAxO* generates a total of 4095 corresponding approximate adders. For assessing the performance of the designs, we have used the $PDP \times LUT$ metric to incorporate all design metrics. The Pareto front analysis in Fig. 5.17 shows that *AppAxO*-generated designs have more hypervolume contribution for all comparisons. For example, Fig. 5.17a shows that *AppAxO*-generated non-dominated points have 1.11% more hypervolume contribution than the ApproxFPGAs non-dominated design points. Further, the

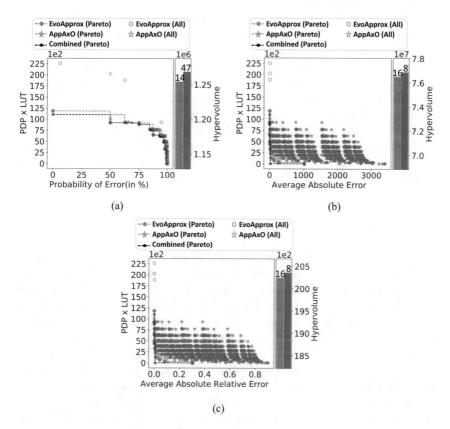

Fig. 5.17 Accuracy and performance comparison of *AppAxO*-generated unsigned adders with the approximate adders from ApproxFPGAs [16]

Table 5.9 Comparison of packing efficiency for implementing a vector adder using 8-bit adders provided by *AppAxO* and ApproxFPGAs [16]

Design	Adder	Single adder LUTs	Average absolute error	Vector adder LUTs	Utilized CLBs
AppAxO	Adder_255	8	0.0	400	50
ApproxFPGAs	add8u_0FP	8	0.0	400	137
AppAxO	Adder_063	6	3.0	300	50
ApproxFPGAs	add8u_0B1	5	2.8	250	91

numbers on the hypervolume bar show the individual non-dominated design points of each library. For example, the Pareto front analysis of only the *AppAxO* designs in Fig. 5.17a shows a total of 47 non-dominated design points. Similarly, the Pareto front analysis of ApproxFPGAs provides 14 non-dominated design points. These results validate the efficacy of the *AppAxO* methodology in generating new approximate designs that can be selected according to an application's accuracy and performance requirements.

Further, the LUTs and carry chains' specific approximation methodology of *AppAxO* enables better packing efficiency of the FPGA resources (CLB utilization) than ApproxFPGAs [16]. For example, to compare the packing efficiency of *AppAxO*-generated 8-bit adders with the 8-bit designs of ApproxFPGAs, we implemented two different versions of a 400-bit vector adder utilizing 50 instances of 8-bit adders. In the first version, we utilized accurate adders, and in the second version, we utilized adders having comparable accuracy. The results of these experiments are shown in Table 5.9. As can be observed that for both experiments, *AppAxO*-generated adders provide better packing efficiency by utilizing a fewer number of CLBs. Similar efficacy of *AppAxO*-generated designs is observed in comparison to other designs of ApproxFPGAs.

5.5.7.2 Proposed Approximate Multipliers

The resulting design points obtained from all explorations related to *AppAxO*, AppAxO_TRN, AppAxO_MBO, and AppAxO_ML were compared with the multipliers proposed in EvoApprox [14]. A total of 3987 8×8 multiplier design configurations were characterized for stand-alone and application-specific performance estimation. Figure 5.18 shows the comparison of the Pareto front obtained for standalone multipliers and approximate multipliers used in the three test applications. It can be observed that the hypervolume of *AppAxO* is lower than that of EvoApprox in the stand-alone multipliers. However, in the search for application-specific multipliers, we report considerable improvements with *AppAxO* for all three applications. If we consider the combination of EvoApprox and *AppAxO*, we observed 10, 18, and 20 Pareto front points with *AppAxO* compared to 8, 1, and 1 points with EvoApprox for ECG, MNIST and GS, respectively.

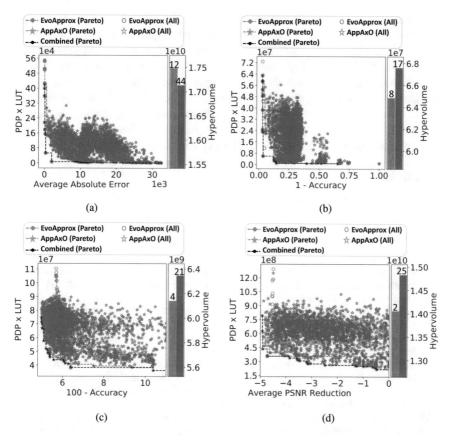

Fig. 5.18 Comparison of the state-of-the-art multipliers in EvoApprox [14] to that of *AppAxO*. The design points for *AppAxO* were obtained across different exploration methods including randomized search, AppAxO_MBO, and AppAxO_ML. (**a**) MULT 8 × 8. (**b**) MNIST. (**c**) GS

5.6 Conclusion

Most state-of-the-art approximate arithmetic operators follow an application-agnostic design methodology. These operators do not have a generic approximation methodology to implement new approximate designs for an application's changing accuracy and performance requirements. We address these limitations in this chapter by presenting the *AppAxO* methodology. *AppAxO* presents a generic methodology for designing approximate operators according to an input configuration defined by an application. The chapter presents the *AppAxO* methodology using approximate adders and multipliers. However, *AppAxO* methodology is generic and can be applied to any arithmetic circuit utilizing 6-input LUT and the carry chains for its implementation. The operator input configuration defines the number of active LUTs involved in the implementation of the operator. For example, for multipliers, the input configuration denotes the LUTs used in partial product generation. *AppAxO* utilizes an MBO-based technique to produce only those operators' configurations that satisfy an application's accuracy and performance constraints. Our methodology also deploys various ML models to use GA to explore the large design space of individual operators and their utilization in various applications by estimating the behavioral accuracy and corresponding performance gains. Compared to state-of-the-art approximate multipliers, *AppAxO* generates non-dominated design points with more hypervolume contribution for various applications. Further, by defining a LUT-carry chain-like building block for ASIC-based systems, *AppAxO* can be extended to design approximate arithmetic operators for ASICs.

References

1. V.K. Chippa, S.T. Chakradhar, K. Roy, A. Raghunathan, Analysis and characterization of inherent application resilience for approximate computing, in *2013 50th ACM/EDAC/IEEE Design Automation Conference (DAC)* (2013), pp. 1–9
2. S. Mittal, A survey of techniques for approximate computing. ACM Comput. Surv. **48**(4), 1–33 (2016)
3. S. Ullah, S.S. Murthy, A. Kumar, SMApproxlib: Library of FPGA-based approximate multipliers, in *2018 DAC* (IEEE, 2018), pp. 1–6
4. S. Ullah, S. Rehman, B.S. Prabakaran, F. Kriebel, M.A. Hanif, M. Shafique, A. Kumar, Area-optimized low-latency approximate multipliers for FPGA-based hardware accelerators, in *Proceedings of the 55th Annual Design Automation Conference*, DAC '18 (Association for Computing Machinery, San Francisco, California, 2018)
5. S. Ullah, H. Schmidl, S. Satyendra Sahoo, S. Rehman, A. Kumar, Area-optimized accurate and approximate softcore signed multiplier architectures. IEEE Trans. Comput. **70**(3), 384–392 (2021)
6. V. Mrazek, R. Hrbacek, Z. Vasicek, L. Sekanina, EvoApprox8b: Library of approximate adders and multipliers for circuit design and benchmarking of approximation methods, in *Design, Automation Test in Europe Conference Exhibition (DATE), 2017* (2017), pp. 258–261
7. M. Shafique, W. Ahmad, R. Hafiz, J. Henkel, A low latency generic accuracy configurable adder, in *Proceedings of the 52nd Annual Design Automation Conference. DAC '15* (Association for Computing Machinery, San Francisco, California, 2015)
8. V. Gupta, D. Mohapatra, A. Raghunathan, K. Roy, Low-power digital signal processing using approximate adders. IEEE Trans. Comput. Aided Des. Integr. Circ. Syst. **32**(1), 124–137 (2013)
9. B.S. Prabakaran, S. Rehman, M.A. Hanif, S. Ullah, G. Mazaheri, A. Kumar, M. Shafique, De-MAS: An efficient design methodology for building approximate adders for FPGA-based systems, in *2018 Design, Automation Test in Europe Conference Exhibition (DATE)* (2018), pp. 917–920
10. V. Mrazek, S.S. Sarwar, L. Sekanina, Z. Vasicek, K. Roy, Design of power-efficient approximate multipliers for approximate artificial neural networks, in *Proceedings of the 35th International Conference on Computer-Aided Design*, ICCAD '16 (Association for Computing Machinery, New York, NY, USA, 2016)
11. Z. Ebrahimi, S. Ullah, A. Kumar, SIMDive: Approximate SIMD soft multiplier-divider for FPGAs with tunable accuracy, in *Proceedings of the 2020 on Great Lakes Symposium on VLSI*, GLSVLSI '20 (Association for Computing Machinery, Virtual Event, China, 2020), pp. 151–156

12. S. Rehman, W. El-Harouni, M. Shafique, A. Kumar, J. Henkel, J. Henkel, Architectural-space exploration of approximate multipliers, in *2016 IEEE/ACMInternational Conference on Computer-Aided Design (ICCAD)* (IEEE, 2016), pp. 1–8

13. P. Kulkarni, P. Gupta, M. Ercegovac, Trading accuracy for power with an underdesigned multiplier architecture, in *2011 24th Internatioal Conference on VLSI Design* (IEEE, 2011), pp. 346–351

14. V. Mrazek, L. Sekanina, Z. Vasicek, Libraries of approximate circuits: Automated design and application in CNN accelerators. IEEE J. Emerg. Sel. Top. Circ. Syst. **10**(4), 406–418 (2020)

15. V. Mrazek, M.A. Hanif, Z. Vasicek, L. Sekanina, M. Shafique, AutoAx: An automatic design space exploration and circuit building methodology utilizing libraries of approximate components, in *Proceedings of the 56th Annual Design Automation Conference 2019*, DAC '19 (Association for Computing Machinery, Las Vegas, NV, USA, 2019)

16. B.S. Prabakaran, V. Mrazek, Z. Vasicek, L. Sekanina, M. Shafique, ApproxFPGAs: Embracing ASIC-based approximate arithmetic components for FPGA-based systems, in *2020 57th ACM/IEEE Design Automation Conference (DAC)* (IEEE, 2020), pp. 1–6

17. A. Liaw, MatthewWiener, et al., Classification and regression by random-forest. R News **2**(3), 18–22 (2002)

18. A.J. Smola, B. Schölkopf, A tutorial on support vector regression. Stat. Comput. **14**(3), 199–222 (2004)

19. L. Bottou, Large-scale machine learning with stochastic gradient descent, in *Proceedings of COMPSTAT'2010* (Springer, 2010), pp. 177–186

20. J.H. Friedman, Stochastic gradient boosting. Comput. Stat. Data Anal. **38**(4), 367–378 (2002)

21. C. Apté, S. Weiss, Data mining with decision trees and decision rules. Future Gener. Comput. Syst. **13**(2-3), 197–210 (1997)

22. F. Murtagh, Multilayer perceptrons for classification and regression. Neurocomputing **2**(5-6), 183–197 (1991)

23. M. Abadi et al., Tensorflow: Asystem for large-scale machine learning, in *12th {USENIX} Symposium on Operating Systems Design and Implementation ({OSDI} 16)* (2016), pp. 265–283

24. F. Biscani, D. Izzo, *esa/pagmo2: pagmo 2.9* (Aug. 2018)

25. J. Pan, W.J. Tompkins, A real-time QRS detection algorithm. IEEE Trans. Biomed. Eng. **BME-32**(3), 230–236 (1985)

26. G.D. Clifford, C. Liu, B. Moody, L.-w.H. Lehman, I. Silva, Q. Li, A.E. Johnson, R.G. Mark, AF classification from a short single lead ECG recording: The PhysioNet/computing in cardiology challenge 2017, in *2017 Computing in Cardiology (CinC)* (2017), pp. 1–4

27. J. Bachrach, H. Vo, B. Richards, Y. Lee, A. Waterman, R. Avižienis, J. Wawrzynek, K. Asanović, Chisel: Constructing hardware in a scala embedded language, in *Proceedings of the 49th Annual Design Automation Conference*, DAC '12 (Association for Computing Machinery, San Francisco, California, 2012), pp. 1216–1225

28. MNIST-cnn, https://github.com/integeruser/MNIST-cnn (2016)

Chapter 6
A Framework for Cross-Layer Approximations

6.1 Introduction

The paradigm of approximate computing has shown promising capabilities for designing energy-efficient computing systems for error-resilient applications. The approximate computing paradigm encompasses different layers of the computation stack. For example, loop perforation [1], precision scaling [2], and utilization of inexact hardware blocks [3] are the commonly investigated techniques at the software, architecture, and circuit levels, respectively. At any layer of the computation stack, the deployed approximation techniques have a limited impact on the resulting output quality of the application and the corresponding performance gains of the implementation. However, depending upon the application, a specific computational layer, or a combination of layers, may prove a better candidate for approximations than other layers in the computation stack. For example, consider a 3×3 convolution kernel sliding over a 5×5 image in Fig. 6.1. The sliding step size of the kernel is defined by the *Stride* parameter of the convolution operation. For each position of the kernel, nine element-wise multiplications followed by an addition are performed to compute a single value of the output image. For *Stride* $= 1$, a total of 81 multiplications are performed to compute a 3×3 output image. As convolution is a commonly used operation in the error-resilient applications such as deep neural networks and image processing, the performance and energy efficiency of the convolution operation can be improved by utilizing approximate multipliers. However, the utilization of only approximate arithmetic units (circuit-level approximation) may not satisfy the resource, performance, and energy budgets of low-power and resource-constrained embedded systems. To improve the overall performance of the convolution operator in Fig. 6.1, it is advantageous to examine other layers of the computation stack for feasible approximations. For example, the total number of energy-consuming multiplication operations, hence the resulting energy consumption, can be reduced by making *Stride* $= 2$ (algorithmic-level

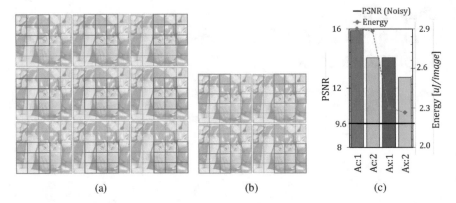

(a)   (b)   (c)

Fig. 6.1 Convolution operation using a 3 × 3 kernel. (a) *Stride* = 1, (b) *Stride* = 2, (c) accuracy/energy trade-off for the Gaussian image smoothing filter using 2 DoFs

approximation). The utilization of approximate multipliers for the reduced number of operations can further improve the energy efficiency of the convolution operation.

Figure 6.1c shows the performance trade-offs from such a *cross-layer approximation* approach by comparing the peak signal-to-noise ratio (PSNR) and energy consumption values of a Gaussian image smoothing filter having a 3 × 3 convolution kernel using accurate (*Ac*) and approximate (*Ax*) multipliers[1] for two different values of the *Stride* parameter. The convolution operation using *Stride* = 1 and accurate multipliers (*Ac*:1) produces an output image with the highest PSNR and energy consumption values. The convolution operation utilizing algorithmic- and circuit-level approximations (*Stride* = 2 with approximate multipliers *Ax*:2) produces the most energy-efficient output, albeit with a low PSNR value. Such a cross-layer approach presents the designer with multiple tuning knobs for application-specific optimizations across multiple Degrees of Freedom (DoFs) in each layer.

However, each DoF and the available choices for it expand the design space exponentially. For instance, increasing the choice of multipliers for each operation, in Fig. 6.1c, from two to three increases the possible design points from 2×2^9 to 2×3^9. Therefore, efficient design space exploration (DSE) frameworks, which enable the joint analysis of multiple DoFs across layers, are necessary to implement cross-layer approximations. Further, the required DSE technique should provide methods for the fast and accurate estimations of the applications' output accuracy and their performance parameters.

The rest of the chapter is organized as follows. Section 6.2 reviews the related state-of-the-art works. Based on the limitations of the state-of-the-art works, we present our novel contributions of the chapter. Section 6.4 presents an error analysis of the approximate arithmetic circuits and presents our novel polynomial regression-

---

[1] mul8s_1KVL from [4] has been used.

based representation of approximate arithmetic circuits. Section 6.5 describes our technique for the performance estimation of approximate accelerators. The MBO-based DSE methodology is discussed in Sect. 6.6. Section 6.7 presents the implementation results of our proposed framework. Finally, Sect. 6.8 concludes the chapter.

6.2 Related Work

Most state-of-the-art works in the domain of approximate computing have focused on harvesting the performance gains by designing and utilizing approximation techniques at a single layer of the design stack. For example, the work presented in [5] utilizes a data sampling technique to process only a subset of input data. The authors of [6, 7] have used reduced precision of data to decrease the application's computational complexity. Techniques presented in [1, 8, 9] utilize loop perforations and task skipping to bargain output accuracy of applications with performance gains. Many works, such as [3, 4, 10], employ approximate arithmetic blocks for realizing energy-efficient hardware accelerators. The work presented in [11] has focused on the circuit-level approximations by performing fast estimations of the optimal design points for implementing area-efficient approximate hardware accelerators.

Some recent related works, such as [1, 12–15], have presented the opportunities offered by the cross-layer approximations. However, most of these works, such as [12], discuss approximation techniques on various layers of computation stack in isolation. These works do not exploit the challenges and opportunities offered by approximations on a combination of layers. The authors of [1] have explored approximations at the algorithm, architecture, and circuit levels of design abstraction to implement a processor, with $1.2x$ to $5x$ energy efficiency, for the recognition and mining (RM) applications. However, their work does not consider utilizing already available open-source approximate arithmetic modules for circuit-level approximation. Further, they do not consider the fast estimations of the feasible design points for their RM processors. The authors of [13] and [14] have proposed simulators for evaluating the impact of three DoF (low bit-width quantization schemes, activations pruning, and approximate multipliers) on the output accuracy of a deep neural network (DNN). The work in [15] has also considered these three DoF for energy-efficient approximate DNNs. However, the state-of-the-art works, summarized in Table 6.1, do not consider the thorough exploration of the design space and fast estimations of the feasible design points provided by various available DoF. Further, to the best of our knowledge, none of the related works provide a solution for analyzing the application-level impact of a new approximate arithmetic unit without the time- and resource-consuming process of actual behavioral (or synthesis) testing of the application.

Table 6.1 Comparing related works

Article	Cross-layer approximation	Operators' modeling and error analysis	Fast estimation of feasible designs	Application analysis of new operators
[5]	✗	✗	✗	✗
[6]	✗	✗	✗	✗
[7]	✗	✗	✗	✗
[1]	✓	✗	✗	✗
[8]	✗	✓	✗	✓
[9]	✗	✗	✗	✗
[4]	✗	✗	✗	✗
[3]	✗	✗	✗	✗
[10]	✗	✗	✗	✗
[16]	✗	✗	✗	✗
[17]	✗	✗	✗	✗
[11]	✗	✗	✓	✗
[12]	✗	✗	✗	✗
[13]	✓	✗	✗	✗
[14]	✓	✗	✗	✗
[15]	✓	✗	✗	✗
[18]	✗	✓	✗	✓
[19]	✗	✓	✗	✗
CLAppED	✓	✓	✓	✓

6.3   Contributions

To address the limitations of the state-of-the-art works, we present an efficient exploration framework, referred to as *CLAppED*, that incorporates a joint analysis of tuning the DoF across multiple layers of the computation stack. Figure 6.2 presents the various stages of the *CLAppED* framework. The related novel contributions are the following:

- We propose a novel PR-based characterization of approximate arithmetic units. Compared to the traditional distribution-based models, we report significant reductions in estimation errors.
- We provide a behavioral framework that utilizes various machine learning models for analyzing the impact of various DoFs on an application's output accuracy. The PR-based coefficients enable machine learning models to correlate an approximate arithmetic operator's impact on an application's output quality.

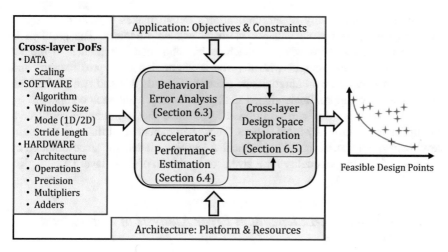

Fig. 6.2 *CLAppED* proposed framework

This behavior allows the trained machine learning models to characterize the application-level impact of new unseen approximate arithmetic units.

- Although primarily focused on behavioral analysis, we present a complete framework for enabling cross-layer approximation-aware DSE. *CLAppED* utilizes behavioral error analysis and accelerator's performance estimates to provide design points that offer better trade-offs between application error and hardware performance.

6.4   Error Analysis of Approximate Arithmetic Units

In this chapter, we have considered only the approximate multipliers to describe the proposed error analysis. However, a similar analysis is also valid for other types of approximate circuits, such as adders and dividers. Further, due to the significance of signed numbers in various modern applications, such as machine learning, we have examined the open-source signed multipliers provided by [4] and [10].[2] Traditionally, statistical error metrics like average relative error, error probability, and mean error distance [20] are used to characterize approximate circuits. However, as observed in [21], the approximate arithmetic units are static nonlinear systems and may violate several fundamental arithmetic principles such as *commutativity* and *associativity*. Further, the utilization of a statistical error metric to estimate an approximate circuit's impact on an application's output accuracy is mostly unknown. This lack of correlation between statistical error metrics and

---

[2] Presented in Chap. 4.

the corresponding quality of an application's output makes it difficult to select an approximate arithmetic unit out of many available choices. The problem is exacerbated in scenarios where a machine learning model is trained to predict an application's final output with a given configuration of approximations and performs poorly on novel approximate circuits. Towards this end, we perform an error analysis of approximate signed multipliers to represent each component by a set of parameters that can be utilized to estimate the approximate result of a multiplier for an arbitrary input dataset. These parameters are utilized in our high-level behavioral framework to train machine learning models for fast estimation of an application's output quality for a given configuration of cross-layer DoF.

6.4.1 Application-Independent Error Analysis of Approximate Multipliers

We have considered *curve fitting* and PR techniques, as shown in Fig. 6.3, to represent each approximate multiplier by a set of unique parameters. The curve fitting-based technique utilizes the nonlinear least-squares method to fit a function 'f' to the results of an approximate multiplier for all input combinations. The function f utilizes a set of parameters to reduce the error between actual approximate products and the fitted results. Due to the various types of approximations in the available multipliers, a single function cannot be fitted to estimate all multipliers accurately. To identify feasible fitting functions, we perform distribution fitting of all approximate multipliers using several data distributions and evaluate their efficacy

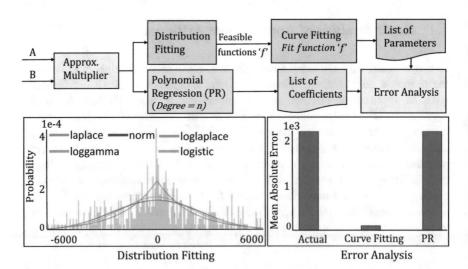

Fig. 6.3 Proposed behavioral error analysis of approximate multipliers with results shown for mul8s_1KR3 approximate multiplier from [4]

using Kolmogorov-Smirnov's (K-S) fitness metric (a commonly used technique to compare a sample with a reference probability distribution) [22]. We select five top distributions to implement the corresponding fitness functions for each approximate multiplier. For example, the distribution fitting subfigure in Fig. 6.3 shows the top five distributions for the multiplier *mul8s_1KR3* from [4]. However, as shown by the Mean Absolute Error (MAE) graph in Fig. 6.3, the fitting of the normal distribution-based function *f* significantly mismatches the actual error value. Similar results with large disparities between actual and estimated statistical error metrics are observed for other multipliers with various *fitting functions*.

The error plot in Fig. 6.3 shows that the PR-based technique can better estimate the approximate results of *mul8s_1KR3*. The PR technique utilizes the *degree* parameter to show the number of coefficients utilized for estimating the target function. For example, for an approximate multiplier with inputs x and y, a degree 2 PR generates six coefficients ($c0$—$c5$), as represented by Eq. 6.1. The fitting of a PR-based model trains these coefficients to minimize the sum of squared errors between actual and estimated outputs for all input combinations of a multiplier. For each approximate multiplier, the PR-based model is trained separately to compute the corresponding coefficients:

$$f(x, y) = c_0 + c_1 x + c_2 y + c_3 x^2 + c_4 xy + c_5 y^2 \qquad (6.1)$$

We have also evaluated the PR-based models' efficacy on the 8-bit approximate adders from [4]. In this regard, we report as low as 18% estimation errors (MAE) with PR-based models compared to 84% estimation errors (MAE) with the curve fitting-based technique.

As shown by the error analysis results in Sect. 6.7, all the trained coefficients, for a particular degree, do not have comparable significance, and the number of actually employed coefficients can be varied to deliver acceptable estimates with a reduced number of trained coefficients. During our error analysis, we have observed that the PR-based technique produces better estimates of the approximate multipliers than the curve fitting-based technique. For example, Fig. 6.4 presents the error distributions (the difference between the actual approximate and the estimated results) of mul8s_1KR3 (a *highly approximate multiplier*) and mul8s_1KVA (a *highly accurate multiplier*) from [4] for all input combinations. For mul8s_1KR3, the logistic-based model[3] produces slightly smaller estimation errors than the norm-based model. For mul8s_1KVA, the norm-based model produces better results than the logistic-based model. However, for both multipliers, the PR-based models produce fewer and smaller estimation errors than the curve fitting-based models. For example, for mul8s_1KVA, the PR-based model generated estimation errors range from −4 to +2. We have observed similar efficacy of the PR-based models for other approximate multipliers. Therefore, we have considered only the PR-based

---

[3] The logistic- and norm-based models for these multipliers have been decided based on the corresponding distribution fitting ranking.

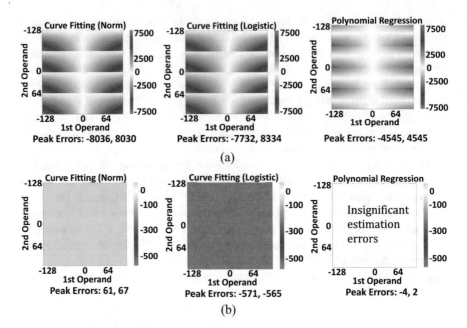

Fig. 6.4 Comparison of estimation-induced errors for multipliers modeled using curve fitting and polynomial regression techniques. (**a**) Approximate multiplier mul8s_1KR3. (**b**) Approximate multiplier mul8s_1KVA

models for application-level error analysis and training of machine learning models for accuracy and performance predictions.

6.4.2 Application-Specific Error Analysis

The behavioral error estimation constitutes the most application-specific analysis stage in *CLAppED*. Although in this chapter, *CLAppED* is tested with one application—*Gaussian smoothing*—the proposed framework is application-agnostic in principle. With appropriate interfaces between the generic DSE method and the application-specific estimation methods, the proposed framework can be used for any arbitrary application. For instance, in the behavioral error estimation for Gaussian smoothing, we implemented a version of the two-dimensional (2D) convolution that integrates the impact of different DoF such as scaling, stride length variation—with and without downsampling—,[4] convolution mode (2D or 1D-Horizontal (1DH) $\rightarrow$ 1D-Vertical (1DV)), and using different approximate

---

[4] Downsampling refers to not using any values for the skipped pixels, leading to reduced output image size. Else, the skipped pixel values are just copied from the input image.

multipliers. Consequently, the impact of any arbitrary configuration across these DoF can be evaluated directly by executing the corresponding executable over a set of images.

Further, we also provide an interface where a supervised ML-based model, which has been trained over a set of randomly generated configurations, can be used to predict the application's output quality for newer cross-layer approximation configurations. In this regard, we use the various available DoF, as shown in Fig. 6.2, to produce input configurations for generating training and testing data for our ML models. A configuration denotes an arbitrary combination of the various DoF. For each configuration in the training and testing set, we implement the applications' behavioral functions to generate corresponding *true labels* (actual outputs). These models can be used for finding the appropriate *feasible* cross-layer approximation-based designs for any arbitrary application.

A similar approach of utilizing supervised ML models for evaluating the impact of various approximate multipliers on the output quality of different applications was presented in Chap. 5. However, the application-level ML models in Chap. 5 have considered a homogeneous accelerator design, i.e., all the approximate multipliers in an arbitrary accelerator configuration were of the same type. For example, for a Gaussian image smoothing filter using a 3×3 kernel, all *nine* multipliers utilized a single approximate design. However, in such a homogeneous accelerator design methodology, accelerator configurations having approximate multipliers (or arithmetic operators) of different types remain unexplored. Towards this end, the proposed *CLAppED* framework provides a more detailed exploration of the accelerator design space by allowing a heterogeneous composition of approximate operators in an accelerator configuration. The *CLAppED* framework provides the coarse-grained and fine-grained arrangement of approximate operators in an application, as illustrated in Fig. 6.5 for ECG application. Figure 6.5a presents the five stages (filters) of the Pan-Tompkins algorithm. In the coarse-grained configuration, all five stages utilize different approximate multipliers. However, all the multipliers in a single stage have the same type of approximation. Figure 6.5b shows an example of a fine-grained approximation where the nine multipliers, in a 3×3 kernel, utilize different approximate multipliers.

Chapter 5 has used an N-bit string to identify the 2^N different approximate designs of an operator. For example, a 36-bit string was used to identify the various 8×8 signed approximate multipliers. The N-bit string identifies each multiplier during the generation of the training and testing datasets for various ML models. Further, the impact of each approximate multiplier on the output quality of the application was evaluated by instantiating a behavioral model of the multiplier. To avoid instantiating approximate multipliers' behavioral functions for each multiplication, we can also create dictionaries of the approximate results for all input combinations and then performing look-ups on the created dictionaries. However, there are two limitations of using the N-bit identifier-based method:

- There is no specific correlation between the N-bit string and the functionality of the corresponding arithmetic operator. For example, the same N-bit string can

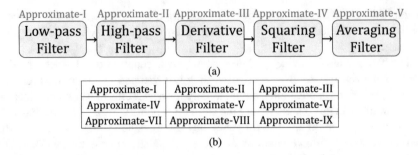

(b)

Fig. 6.5 Heterogeneous composition of approximate operators supported by *CLAppED*. (**a**) Coarse-grained heterogeneity: different filters utilizing different approximate arithmetic units. (**b**) Fine-grained heterogeneity: a single filter utilizing different approximate arithmetic units for each operation

identify either an approximate multiplier or an approximate operator. The lack of correlation between the N-bit identifier and the corresponding function can limit any algorithm's ability in general, particularly the ML models, to understand the operator's functionality and evaluate its impact.

• The instantiation of approximate multipliers' behavioral models for performing multiplication is a computationally intensive operation. For example, the state-of-the-art DNNs employ millions of multiplication operations to infer a single image. Therefore, the behavioral exploration of the heterogeneous combinations of approximate multipliers for these DNNs can significantly increase the overall exploration time.

To resolve these two limitations, the *CLAppED* framework utilizes the PR coefficient to identify and implement the functionality of each approximate operator. As described in Eq. 6.1, the PR technique utilizes trained coefficients to compute the functionality of an operator using a generic equation. Therefore, the values of the trained PR coefficients define the functionality of the operator. The association between the values of the coefficients and the behavior of the approximate operator also helps ML models to evaluate their impact on the output quality of an application.

6.5   Accelerator Performance Estimation

Accelerator performance estimation constitutes the architecture-specific design method of *CLAppED*. Figure 6.6 shows the various stages in this method. Stage ① involves designing the various accelerators that can be used for the application. This stage should consider the effect of only those DoF that can result in different

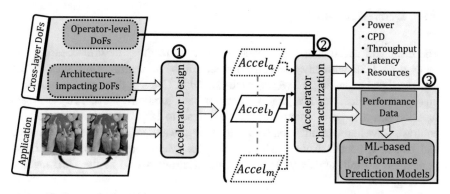

Fig. 6.6 Accelerator performance estimation

accelerator architectures and not the operator-level approximations. For instance, in the example of 2D convolution for Gaussian smoothing, varying the window size or the mode of convolution results in different numbers and types of accelerators. To present *CLAppED* framework, we implemented the line buffer-based sliding window [23] accelerators for 2D, 1DH, and 1DV operations. Further, the HLS-based designs support the variation in odd-numbered window sizes and different stride lengths, resulting in varying accelerator designs. The second stage, ②, considers the effect of using operator-specific approximations in the accelerator. For our current work, we include the accelerator characterization with approximate signed multipliers from [4, 10]. It must be noted that, in this work, we do not use the proposed framework for exploring the impact of HLS directives. We limit the exploration to configuring the approximate multipliers and other application-specific DoF. Prior works such as [24, 25] present exploration methodologies for HLS and can be used alongside ours of the approximate operations.

While the accurate characterization of the accelerators with varying *DoFs* provides high-quality results, the *actual* performance estimation using synthesis tools is very time-consuming. For instance, the characterization of a 2D convolution accelerator of 3×3 window size takes around 15 min. To this end, stage ③ in Fig. 6.6 shows an ML-based performance prediction method to estimate the hardware metrics. For 2D convolution, the dimensions of the model for each performance metric are a subset of the following features: the input image size, the stride length, a downsampling flag, and the precision and type of each multiplier operation in the design. Similar to the behavioral error analysis, the actual and the ML-based approaches provide alternative methods of varying result quality and estimation time to the DSE methodology of *CLAppED*.

6.6   DSE Methodology

The error and accelerator performance estimation methods described earlier can be used for DSE with any randomized algorithm-based optimization, such as GA and simulated annealing. However, as discussed in Chap. 2, these methods usually involve the evaluation of a large number of configurations. Hence, if the fitness function has a large evaluation time, it can lead to large DSE times as well. In contrast, Bayesian Optimization (BO) [26] provides a more directed search method and can also benefit from faster design point evaluations. In this chapter, we implement a Multi-objective Bayesian Optimization (MBO) as the DSE method in *CLAppED*.

In this chapter, we model the cross-layer approximation design as an optimization problem and implement a framework that allows the novel ML-based estimation methods to be used as low-cost fitness functions required for faster DSE. Towards this end, we generate multiple probabilistic models, one for each design objective. For instance, in the joint optimization for application accuracy and the LUT utilization, the *surrogate function* consisted of separate probabilistic models (Gaussian process regression) for both the objectives. The implementation of the *acquisition function* involved generating random cross-layer approximation configurations, followed by predicting their objective metrics with the corresponding probabilistic models in the *surrogate function* and ranking the samples based on their exclusive hypervolume contributions. A fixed number of the top-ranked samples are added to the configurations set for the next iteration.

The previous two sections present both *true* and ML-based estimation methods. The *true* estimation involved reporting the metrics from actual implementation of an accelerator with the cross-layer approximation configurations. Both *true* and ML-based methods serve as alternatives for the *objective function* only. Although the *surrogate function* is also based on a probabilistic model, unlike the ones used in the *objective function*, it is independent of the application under test. The proposed framework allows the designer to choose either of the methods (*true*/ML-based) independently for error and hardware performance estimation during *objective function* evaluation. In case of using ML-based methods, the DSE results can be used in a neighborhood search with the *actual* evaluation to further improve the quality of results.

6.7   Results and Discussion

6.7.1   Experimental Setup and Tool Flow

The implementation of *CLAppED* involved both probabilistic analysis and hardware design. The HLS-based accelerator designs were implemented in C++ and synthesized with Xilinx Vivado Design Suite. All designs have been implemented for

Xilinx Zynq UltraScale+ MPSoC (xczu3eg-sbva484-1-e device). The probabilistic analysis for curve fitting, Polynomial Regression (PR), application-level ML-based modeling, and the MBO-based DSE methodology was implemented in Python using multiple packages, including scikit-learn, TensorFlow [27], PyGMO [28].

6.7.2 Behavioral Analysis

A PR model's efficacy to predict an approximate multiplier's output is computed using the model's coefficient of determination denoted as R^2. A large value of R^2 (≈ 1) denotes a good fitting of the model to predict the actual outputs. As shown in Eq. 6.1, the complexity of a PR model is defined by its degree parameter. During the error analysis of the approximate multipliers from [4] and [10], we have observed that PR models with at least degree 3 produce significantly accurate estimations of the actual approximate results. Further, all the generated coefficients for a PR model (after training the model) do not have equal significance. We can analyze the generated coefficients of all multipliers, under consideration, for a specific degree-based PR model and rank their overall significance. Based on the ranking of coefficients, we can remove the less significant coefficient for each multiplier. For example, Fig. 6.7 compares the *actual* and *estimated* (using degree 3 PR models) average absolute relative errors of seven approximate multipliers from [4]. The *Clipped_8*, *Clipped_6*, and *Clipped_5* bars in the figure show the utilization of only 8, 6, and 5 trained coefficients for PR models to estimate approximate multipliers. The PR models provide significantly accurate estimates of the actual average relative error values, on average a difference of 15%. The *Clipped_5* graphs show, on an average, only a 0.06% degradation in the estimated values for the multipliers under consideration.

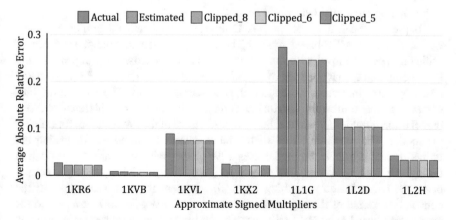

Fig. 6.7 Error analysis of various approximate signed multipliers from EvoApprox signed multipliers library [4]

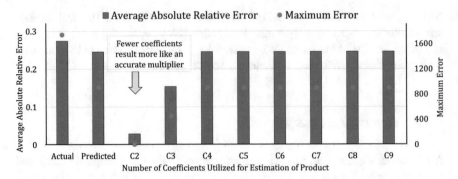

Fig. 6.8 Error analysis of mul8s_1L1G approximate multiplier from EvoApprox signed multipliers library [4]

Utilizing the trained coefficients' ranking, we can also implement custom PR models that are *retrained* with a reduced number of coefficients. For example, Fig. 6.8 compares the average relative and maximum errors of *mul8s_1L1G*, from [4], using different numbers of coefficient (C2–C9)-based retraining. The *predicted* bar in the figure shows the estimation utilizing all the trained coefficients. As shown by the results, PR models with only 2 and 3 coefficients (C2 and C3, respectively) produce large deviations from the corresponding approximate results. Due to the reduced number of coefficients, these models behave like accurate multipliers. However, PR models having more than three coefficients produce high R^2 values and efficiently represent the approximate multiplier. For example, for the actual average relative error (1.36) and maximum absolute error (8001), a C4-based PR model produces 1.34 and 5012 for these error metrics, respectively. Further, increasing the number of coefficients beyond 6 has no impact on the PR model's accuracy for *mul8s_1L1G*. The fine-grained control on utilizing an only appropriate number of coefficients also helps in reducing the complexity of ML models for estimating applications' behavioral accuracy.

To predict the impact of various DoF on an application's output accuracy, we have used a MLP-based model. The model is used to predict accuracy of achieving noise removal with random configurations of cross-layer approximation DoF, compared to that obtained by a *golden* configuration. The model is trained and tested by providing an input dataset of 2000 configurations. These configurations contain various uniformly distributed combinations of the considered DoF. Our model utilizes randomly selected 80% configurations of the input dataset for training the model. The remaining 20% of the input dataset is employed for testing the trained model. Further, the model uses 20% of the training dataset for validation during the training phase. For this work, we have modeled a 3 × 3 kernel-based Gaussian image smoothing filter. To represent the nine multipliers in the convolution kernel of the application, we have used *index-*, *error metrics-*, and PR coefficients-based methods. In the *index*-based method, each multiplier is assigned a unique random value for its identification. In the *error metrics*-based method,

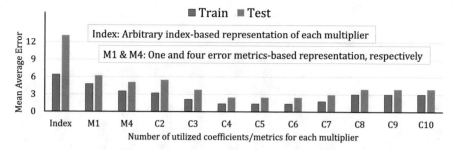

Fig. 6.9 Mean Absolute Error (MAE) of training and testing an MLP for a 3 × 3 kernel-based Gaussian image moothing filter utilizing different numbers of coefficients for multipliers from [4] and [10]

we have utilized each multiplier's four statistical error metrics (maximum absolute error, average relative error, error probability, and mean squared error) for its identification. We refer to this configuration as *M*4 in our experiments. We have also experimented with using only one single statistical error metric, referred to as *M*1, for the identification of multipliers in the MLP model. Such single values-based identification is used in [11], which utilizes the weighted mean error distance (WMED) of an adder for its identification. However, our application-level error analysis reveals that *PR coefficients*-based representation of a multiplier performs better than other techniques.

Figure 6.9 compares the training and testing accuracy of the utilized MLP model. The *M*1-based representation shows the mean squared error-based identification. The *C2–C10* shows the number of utilized PR coefficients to represent each multiplier in the training and testing datasets. The index-based method produces the highest Mean Absolute Error (MAE) for both the training and the testing phases, 6.37 and 13.12, respectively. Clearly, the model cannot identify the correlation between multipliers' indices and the impact of utilized approximate multipliers on the output quality degradation. The *M*1- and *M*4-based techniques produce lower MAE values than the index-based method. For example, *M*4-based representation produces 3.5 and 5.0 MAE values for training and testing datasets, respectively. However, the utilization of PR coefficients-based representation improves the MLP model's accuracy significantly. For example, with even *two* PR coefficients-based representation, the MLP model's training and testing accuracy improved by having only 3.2 and 5.46 mean average errors, respectively. The *C4*-, *C5*-, and *C6*-based multipliers' representation offers the minimum average error of the MLP model. For example, the *C4*-based representation (four PR coefficients) generates only 1.4 and 2.4 MAE values for training and testing datasets, respectively. However, increasing the number of coefficients to represent a multiplier directly impacts the total number of trainable parameters of the MLP model. For an MLP model with a high number of trainable parameters, a large dataset is required to train the model. For the given dataset, the *C7*- to *C10*-based representation of multipliers results in reducing the output accuracy of the model due to insufficient training data.

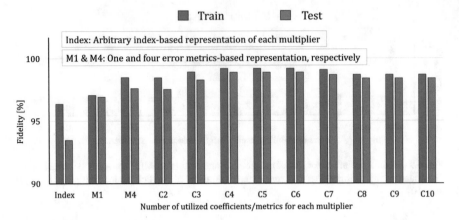

Fig. 6.10 Fidelity of training and testing an MLP for a 3 × 3 kernel-based Gaussian image smoothing filter utilizing different number of coefficients for multipliers from [4] and [10]

To further evaluate the performance of the MLP model, we have used the *fidelity* metric [11]. The fidelity metric denotes the relationship ($=, <, >$) between the input configurations and their corresponding actual and predicted values. Figure 6.10 shows the percentage fidelity of the MLP model on training and testing sets using various representations for the multipliers. As previously observed in Fig. 6.9 for MAE analysis, the index-based method produces the minimum fidelity. The *C4*- to *C6*-based representations of multipliers produce the best training and testing fidelity, 99.2% and 98.9%, respectively.

To understand the relationship between the number of PR coefficients and the corresponding inference time of the MLP model, Fig. 6.11a compares the MAE and average inference time for 1000 iterations of the testing dataset. On average, the MAE decreases, and the inference time increases by utilizing more coefficients to represent multipliers in the MLP model. For the given MLP model and the testing dataset, *C4* (employing four PR coefficients to show a multiplier) produces the best result in terms of low MAE and inference time.

The utilization of PR coefficients to represent multipliers enables the MLP model to correlate approximate multipliers and their impact on an application's output accuracy. This correlation enables the MLP model to predict an application's output quality for new multipliers (not included in the training and validation datasets). Figure 6.11b shows the training and testing MAEs of an MLP model for two cases. In the first experiment, the testing dataset configurations also include a multiplier that is not present in any configuration of the training dataset. In the second experiment, the testing dataset has two such multipliers that are not available to any configuration of the training dataset. As shown by the results, the PR coefficients-based representation of the approximate multipliers has enabled the MLP model to produce high output accuracy even for unseen new multipliers (with up to 98.4% fidelity).

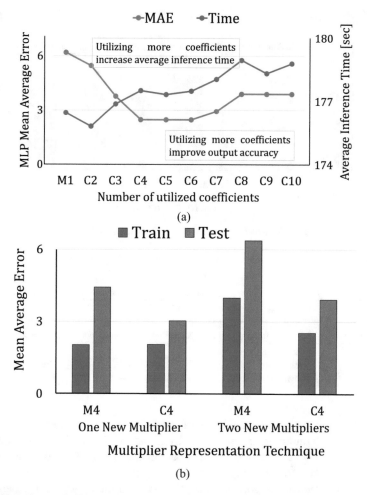

Fig. 6.11 Efficacy of MLP-based estimation of a Gaussian image smoothing filter. (**a**) Impact of utilizing different number of coefficients on the MAE and average estimation time. (**b**) Estimation accuracy on test data having new unseen multipliers

6.7.3 Accelerator Performance Estimation

Similar to the behavioral analysis, we implemented MLP-based models for predicting the performance of accelerators. Figure 6.12 shows the results from the experiments for a 2D convolution accelerator. The bar graphs in the figure show the prediction accuracy, in terms of fidelity of results, for four different metrics—Power-Delay Product (PDP), latency (the number of clock cycles needed for the 2D convolution over an image), the FPGA's LUT utilization, and the accelerator's power dissipation. 1000 designs were used for training the MLP models and 200 design points were used for the testing.

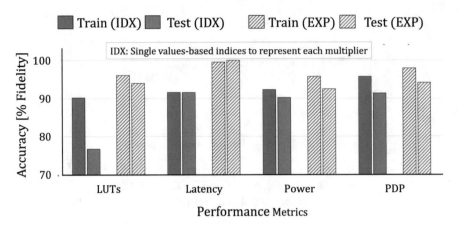

Fig. 6.12 Accelerator performance estimation using MLP-based model

Table 6.2 MLP dimensions for accelerator performance modeling

Performance metric→	PDP	LUTs	Latency	Power dissipation
Accelerator dimensions	Image size, stride length, downsampling	Size, stride length, downsampling	Image size	Image size, stride length, downsampling
Multiplier dimensions	Critical path delay, total power dissipation	LUT utilization	–	Signal power, logic power

The accuracy on training and testing datasets is reported for two approaches—
IDX, where each multiplier is represented by an index value, and *EXP*, an expanded
representation, where we tested different combinations of DoF for determining the
dimensions of the corresponding MLP model. Table 6.2 shows the resulting dimensions for each model, categorized as accelerator- and multiplier-level dimensions.
The set of dimensions for the *EXP*-based model was chosen based on its testing
accuracy. It can be noted from the table that the model for latency has the least
parameters with the highest prediction accuracy. Since latency depends primarily
on the image size, multiplier-specific parameters were not used in the model. It can
be observed that using the *EXP*-based approach leads to higher accuracy for both
training and testing datasets.

6.7.4 DSE Performance

The *directed* search capability of Bayesian Optimization (BO) is especially beneficial for problems with costly fitness function evaluation—similar to the *actual*
estimation of accelerator performance. Although we provide a faster ML-based

approach for the same, the search capability of the optimization algorithm determines the quality of results. Figure 6.13a compares the progress of a similar search for application-level error and LUT utilization trade-offs, using MBO and random search. The plot shows the hypervolume obtained with an increasing number of design point evaluations. Both methods use the ML-based estimation of application-level error and LUT utilization. It can be observed that using MBO obtains better quality results much faster. The data in the figure is obtained from an optimization run that evaluates ten new samples in each iteration, selected from 50 random samples that are generated by the *acquisition function*.

Figure 6.13b shows the results from the analysis of the design points obtained from MBO-based DSE. In this search for Error-LUT trade-offs, the MBO-based search using MLP models generated 23 Pareto front design points, shown as MBO_MLP_PARETO in the figure. It must be noted that only one among those 23 points, shown as Ⓐ, used a configuration where all the 9 multipliers (for a 3×3 window) were of the same type. This reinforces our motivation of the need for searching among the vast number of possible multiplier permutations in an application. Further, 3 points had a stride length of 2 compared to 1 for the others, and 12 points had downsampling enabled. Similarly, image scaling values 3, 2, and 1 were observed for 2, 19, and 2 design points, respectively. These results further demonstrate the need for cross-layer exploration across multiple types of DoF. We evaluated the 23 points with actual hardware synthesis, and the resulting points are plotted in Fig. 6.13b (ACTUAL_EVAL). It can be observed that the *true* points are close to those obtained using the MLP-based model.

6.8 Conclusion

Approximate computing presents an attractive path for achieving low-cost execution for a large variety of applications. Further, a cross-layer approach allows the design of application-specific approximations with more availability of DoF. In this chapter, we present *CLAppED*, a framework for enabling such a design approach. Our framework utilizes a Polynomial Regression (PR)-based representation of approximate operators. PR-based representations enable ML models to better correlate an approximate operator's coefficients with their impact on the output quality of an application. The proposed behavioral analysis allows the designer to create application-specific models for exploring existing arithmetic operators (with up to 98.9%) and predicting new unseen ones' efficacy (with up to 98.42% fidelity). Further, the fast accelerator performance estimation methods and the MBO-based DSE methodology enable the search for high-quality designs.

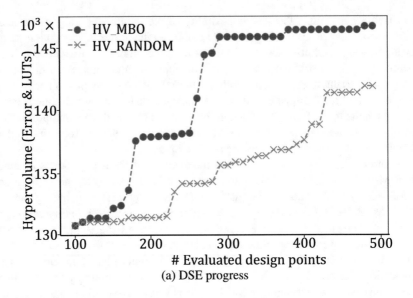

(a) DSE progress

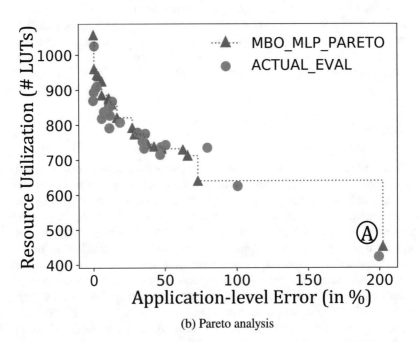

(b) Pareto analysis

Fig. 6.13 (**a**) Comparison of the quality of results (hypervolume) during DSE for trade-offs between application-level error and LUT utilization using MBO and randomized search. (**b**) Analysis of the Pareto points obtained from MBO-based DSE

References

1. V.K. Chippa, D. Mohapatra, K. Roy, S.T. Chakradhar, A. Raghunathan, Scalable effort hardware design. IEEE Trans. Very Large Scale Integr. (VLSI) Syst. **22**(9), 2004–2016 (2014)
2. P. Düben, Parishkrati, S. Yenugula, J. Augustine, K. Palem, J. Schlachter, C. Enz, T.N. Palmer, Opportunities for energy efficient computing: A study of inexact general purpose processors for high-performance and big-data applications, in *2015 DATE* (2015), pp. 764–769
3. S. Ullah, S.S. Murthy, A. Kumar, SMApproxlib: Library of FPGA-based approximate multipliers, in *2018 DAC* (IEEE, 2018), pp. 1–6
4. V. Mrazek, R. Hrbacek, Z. Vasicek, L. Sekanina, EvoApprox8b: Library of approximate adders and multipliers for circuit design and benchmarking of approximation methods, in *Design, Automation Test in Europe Conference Exhibition (DATE), 2017* (2017), pp. 258–261
5. I. Goiri, R. Bianchini, S. Nagarakatte, T.D. Nguyen, ApproxHadoop: Bringing approximations to mapreduce frameworks. SIGPLAN Not **50**(4), 383–397 (2015)
6. S. Vogel, J. Springer, A. Guntoro, G. Ascheid, Selfsupervised quantization of pre-trained neural networks for multiplierless acceleration, in *2019 Design, Automation & Test in Europe Conference & Exhibition (DATE)* (IEEE, 2019), pp. 1094–1099
7. S. Vogel, M. Liang, A. Guntoro, W. Stechele, G. Ascheid, Efficient hardware acceleration of CNNs using logarithmic data representation with arbitrary log-base, in *Proceedings of the International Conference on Computer-Aided Design*, ICCAD '18 (Association for Computing Machinery, San Diego, California, 2018)
8. V.K. Chippa, S.T. Chakradhar, K. Roy, A. Raghunathan, Analysis and characterization of inherent application resilience for approximate computing, in *2013 50th ACM/EDAC/IEEE Design Automation Conference (DAC)* (2013), pp. 1–9
9. S. De, S. Mohamed, K. Bimpisidis, D. Goswami, T. Basten, H. Corporaal, Approximation trade offs in an image-based control system, in *2020 Design, Automation Test in Europe Conference Exhibition (DATE)* (2020), pp. 1680–1685
10. S. Ullah, H. Schmidl, S. Satyendra Sahoo, S. Rehman, A. Kumar, Area-optimized accurate and approximate softcore signed multiplier architectures. IEEE Trans. Comput. **70**(3), 384–392 (2021)
11. V. Mrazek, M.A. Hanif, Z. Vasicek, L. Sekanina, M. Shafique, AutoAx: An automatic design space exploration and circuit building methodology utilizing libraries of approximate components, in *Proceedings of the 56th Annual Design Automation Conference 2019*, DAC '19 (Association for Computing Machinery, Las Vegas, NV, USA, 2019)
12. M. Shafique, R. Hafiz, S. Rehman, W. El-Harouni, J. Henkel, Invited: Cross-layer approximate computing: From logic to architectures, in *2016 53nd ACM/EDAC/IEEE Design Automation Conference (DAC)* (2016), pp. 1–6
13. C. De la Parra, A. Guntoro, A. Kumar, Improving approximate neural networks for perception tasks through specialized optimization. Future Gener. Comput. Syst. **113**, 597–606 (2020)
14. Y. Fan, X. Wu, J. Dong, Z. Qi, AxDNN: Towards the cross-layer design of approximate DNNs, in *Proceedings of the 24th Asia and South Pacific Design Automation Conference*, ASPDAC '19 (Association for Computing Machinery, Tokyo, Japan, 2019), pp. 317–322
15. M. Hanif, A. Marchisio, T. Arif, R. Hafiz, S. Rehman, M. Shafique, X-DNNs: Systematic cross-layer approximations for energy-efficient deep neural networks. J. Low Power Electron. **14**, 520–534 (2018)
16. P. Kulkarni, P. Gupta, M. Ercegovac, Trading accuracy for power with an underdesigned multiplier architecture, in *2011 24th Internatioal Conference on VLSI Design* (IEEE, 2011), pp. 346–351
17. S. Rehman, W. El-Harouni, M. Shafique, A. Kumar, J. Henkel, J. Henkel, Architectural-space exploration of approximate multipliers, in *2016 IEEE/ACMInternational Conference on Computer-Aided Design (ICCAD)* (IEEE, 2016), pp. 1–8
18. S. Mazahir, O. Hasan, R. Hafiz, M. Shafique, Probabilistic error analysis of approximate recursive multipliers. IEEE Trans. Comput. **66**, 1982–1990 (2017)

19. M.K. Ayub, O. Hasan, M. Shafique, Statistical error analysis for low power approximate adders, in *Proceedings of the 54th Annual Design Automation Conference 2017*, DAC '17 (Association for Computing Machinery, Austin, TX, USA, 2017)
20. J. Liang, J. Han, F. Lombardi, New metrics for the reliability of approximate and probabilistic adders. IEEE Trans. Comput. **62**(9), 1760–1771 (2013)
21. M. Bruestel, A. Kumar, Accounting for systematic errors in approximate computing, in *Design, Automation Test in Europe Conference Exhibition (DATE), 2017* (2017), pp. 298–301
22. F.J. Massey Jr., The Kolmogorov-Smirnov test for goodness of fit. J. Am. Stat. Assoc. **46**(253), 68–78 (1951)
23. Xilinx, UG1270: *Vivado HLS Optimization Methodology Guide* (Apr. 2018)
24. H.-Y. Liu, L.P. Carloni, On learning-based methods for design-space exploration with high-level synthesis, in *Proceedings of the 50th Annual Design Automation Conference*, DAC '13 (Association for Computing Machinery, Austin, Texas, 2013)
25. A. Mehrabi, A. Manocha, B.C. Lee, D.J. Sorin, Prospector: Synthesizing efficient accelerators via statistical learning, in *Proceedings of the 23rd Conference on Design, Automation and Test in Europe*, DATE '20 (EDA Consortium, San Jose, CA, USA, 2020), pp. 151–156
26. J. Močkus, On Bayesian methods for seeking the extremum, in *Optimization Techniques IFIP Technical Conference Novosibirsk, July 1–7, 1974*, ed. by G.I. Marchuk (Springer, Berlin, Heidelberg, 1975), pp. 400–404
27. M. Abadi et al., Tensorflow: Asystem for large-scale machine learning, in *12th {USENIX} Symposium on Operating Systems Design and Implementation ({OSDI} 16)* (2016), pp. 265–283
28. F. Biscani, D. Izzo, *esa/pagmo2: pagmo 2.9* (Aug. 2018)

Chapter 7
Conclusions and Future Work

7.1 Conclusions

The last few decades have seen tremendous growth in the computational power of computing systems. The low-cost availability of high-performance computing systems has resulted in the boom and widespread adoption of many computationally intensive applications. For example, today, many mobile devices and embedded systems at the edge provide dedicated processors for neural networks acceleration. This rapid growth in the computational power of the systems has been mainly driven by advancements in technology scaling, and various innovative architectures such as multicore and heterogeneous computing systems.

During the evolution of the computing systems, as mentioned above, the computing industry has tried to maintain a strict notion of computational accuracy across the various layers of abstraction. In these "accurate systems," every computation is performed with maximum available accuracy. However, today, a wide range of applications—from data centers to embedded systems—neither require 100% accurate computations nor produce a single golden answer. These applications possess an inherent error resilience to the inaccuracies (approximations) in their utilized data and employed computations. For these applications, the paradigm of approximate computing can be utilized to implement high-performance and energy-efficient computing systems. The approximate computing paradigm employs the output quality as another design metric (along with resource utilization, performance, and energy consumption) to design computing systems. It covers all layers of the computation stack, and to this end, Chap. 1 in the book summarizes some of the recent works related to approximate computing paradigm.

Most of the state-of-the-art works related to the approximate computing paradigm have considered ASIC-based systems. Our experiments and analysis reveal that the utilization of ASIC-based approximation techniques for FPGA-based systems does not provide comparable performance gains. The main reason for this incomparable performance gain is the architectural differences between ASIC- and

S. Ullah, A. Kumar, *Approximate Arithmetic Circuit Architectures for FPGA-based Systems*, https://doi.org/10.1007/978-3-031-21294-9_7

FPGA-based implementations. The widespread utilization of FPGAs to provide acceleration for various applications, particularly for error-resilient applications, calls for designing FPGA-specific approximation techniques. This book focuses on the design and analysis of approximate arithmetic modules optimized for FPGA-based systems. As multiplication is one of the most used arithmetic operations in various error-resilient applications, such as machine learning and signal processing, the book has mainly focused on the designs of approximate multipliers.

This book approaches the challenge of designing approximate operators by first targeting for an accurate resource-efficient and high-performance baseline architecture. The availability of accurate operators is also important because not all suboperations of an error-resilient application are error-tolerant. To this end, Chap. 3 utilizes 6-input LUTs and associated carry chains of Xilinx FPGAs to propose various designs of accurate unsigned, signed, and constant coefficient-based multipliers. Our unsigned and signed multipliers fuse the generation and accumulation of two consecutive partial products into one step, which results in the reduction of total utilized LUTs and critical path delay of the implementation. For signed multiplication, we have explored Booth's and Baugh-Wooley's multiplication algorithms. The constant coefficient-based multipliers are based on the shift and add method. For this purpose, our implementation analyzes the constant coefficient and utilizes either addition or subtraction to compute the product in a minimum number of steps. The chapter also presents the performance impact of utilizing proposed designs in various applications.

Based on our accurate designs, we present various unsigned and signed approximate multipliers in Chap. 4. These designs trade the multiplication output accuracy for reducing the total number of utilized LUTs, critical path delay, and energy consumption of the multiplier. For unsigned multiplication, we present three 4×4 approximate multipliers and utilize the modular approach for designing higher-order multipliers. For this purpose, the book also contributes by proposing an approximate ternary adder—3 : 1 compressor. The modular approach allows the realization of a large design space of multipliers offering different trade-offs between accuracy and performance. The presented approximate signed multiplier is an array-based design that employs Booth's multiplication algorithm and approximately computes the least significant partial product terms in every partial product row. The chapter also utilizes various error-resilient applications to evaluate the accuracy-performance impact of the proposed multipliers.

Our analysis of employing approximate multipliers in various applications to evaluate their impact on the application-level accuracy and performance reveals that a single approximate design may not be feasible for a diverse range of error-resilient applications. Towards this end, Chap. 5 presents a generic methodology for implementing application-specific approximate operators, such as adders and multipliers, from their corresponding accurate implementations. Our proposed methodology can generate *hundreds* of corresponding approximate designs for a given accurate implementation of an operator. We also provide a Multi-objective Bayesian Optimization (MBO)- and Genetic Algorithm (GA)-based exploration methodology to identify implementations of an approximate operator that can satisfy an application's accuracy and performance constraints. The efficacy evaluation of our proposed methodology on three different benchmark applications shows that it can generate more non-dominated accelerator design configurations than those provided by state-of-the-art designs.

The book also explores other layers of the computation stack to satisfy the accuracy-performance constraints of error-resilient applications. In this regard, to explore the collective impact of approximations at various layers of the computation stack (*cross-layer approximation*), Chap. 6 presents an MBO-based framework. The proposed framework utilizes Multi-layer Perceptrons (MLPs) to estimate the impact of approximations at various layers of the computation stack. A key challenge in utilizing Machine Learning (ML) models for accuracy-performance trade-off estimation is the representation of approximate arithmetic operators in ML models. To this end, the book contributes by proposing a Polynomial Regression (PR)-based representation of approximate arithmetic operators. The proposed PR-based technique enables the ML models to predict the impact of approximate operators with higher accuracy than other techniques.

7.2 Future Works

This book addresses some of the open research challenges related to the design of FPGA-based approximate hardware accelerators for error-resilient applications. However, there are some related research problems that provide an opportunity for a simple and robust design of approximate hardware accelerators.

- *Design and exploration of higher-level approximate operators:* Higher-level operators, such as Multiply-Accumulate (MAC) units and vector multipliers, are commonly utilized operations in various error-resilient applications. Most of these operations can be implemented by employing various combinations of accurate/approximate multipliers and adders. However, in this technique, the operator-specific degrees of freedom for approximation are not truly exploited. For example, the dot product of two vectors with dimensions [1, 3] would require three multiplication and two addition operations to compute the result. However, an approximate dot product operator can also explore other degrees of freedom, such as sampling of the operands and utilization of spatial correlation between individual values of operands to reduce the number of different suboperations and overall computational complexity of the operator. The design of higher-level approximate operators can significantly help implement hardware accelerators for resource-constrained embedded systems.
- *Cross-architecture exploration of accurate and approximate designs:* The various accurate and approximate designs presented in this book have been implemented for Xilinx FPGAs. One of the main reasons for selecting Xilinx FPGAs is their widespread utilization in both academia and industry. Our proposed designs are based on the efficient utilization of 6-input LUTs and the associated carry chains

of the Xilinx FPGAs. Intel FPGAs—one of the highest market shareholders after Xilinx—also support 6-input LUTs and associated adders to implement various types of logic. The adaptive logic module (ALM) in an Intel FPGA receives 8-input and allows the configuration of the associated LUT in various configurations (as we have done for the 6-input LUT in Xilinx FPGAs). A generic framework that can implement accurate/approximate designs for FPGAs from both vendors by utilizing their respective LUT-level primitives can further increase the scope of FPGA-based hardware accelerators.

- *Application-specific approximate operators for ASIC-based systems:* Chap. 5 presented a framework for application-specific approximate operator generation for FPGA-based systems. The framework employed MBO- and GA-based optimization techniques to generate operator configurations. These configurations denoted the logic blocks (a LUT and a carry chain cell) that should be truncated from the accurate implementation of an operator. A similar framework can be implemented for ASIC-based systems by defining a comparable basic logic block in ASIC-optimized operators.

- *Run-time adaptation of approximate designs:* There are some scenarios where the accuracy-performance requirements of a system change at run-time. For example, a battery-operated accelerator should be able to save battery life by switching to the energy-efficient mode that may reduce the generated output quality. Similarly, an accelerator may switch between two operational modes to support different throughput and output accuracy trade-offs. One of the possible solutions to implement such run-time reconfigurability is to support variable precision-based operations. For example, an accelerator may utilize 8-bit operations to produce higher-quality outputs, and it may turn to 4-bit operation mode to provide relatively lower-quality outputs but with better energy efficiency and higher throughput. In the 4-bit operation mode, the same 8-bit operator should support multiple instances of 4-bit operators to provide higher throughput. The accurate and approximate designed proposed in this book are design-time reconfigurable. To this end, there is a need for approximate operator architectures that can be reconfigured at run-time. The modular implementation of operators can be a good starting point in this direction.

- *Polynomial regression-based modeling of higher-level operators:* Chap. 6 utilized the Polynomial Regression (PR) coefficient to model approximate operators such as adders and multipliers. However, can the PR coefficients technique be utilized to model higher-level operators implemented using suboperators? For example, can an approximate MAC operator, implemented using approximate multipliers and adders, be represented using PR coefficients? Is there any relationship between an approximate MAC operator's PR coefficients and its constituent operators' PR coefficients? The answers to these questions can help in the efficient design space exploration of complex error-resilient applications such as ANNs, which utilize such types of higher-level operators.

Index

Printed in the United States
by Baker & Taylor Publisher Services